Take the Pain Out of Pre-Algebra

Take the Pain Out of Pre-Algebra

Success Strategies for Struggling Math Students & Other Kids Too

Important stuff they didn't tell you in your Ed classes
& things you can use on Monday morning.

James Slosson

ISBN: Hardcover 978-1-6698-4035-0
Softcover 978-1-6698-4034-3
eBook 978-1-6698-4033-6

Print information available on the last page.

Rev. date: 08/02/2022

To order additional copies of this book, contact:
Xlibris
844-714-8691
www.Xlibris.com
Orders@Xlibris.com
842457

CONTENTS

Appendices – Extra Stuff

Dedication

For Karen Eitreim who left us way too soon.

Karen was a world traveler, a teacher, a principal, and a friend.

I'm trying to keep this book informal. It's not a scholarly work; I guess it's more like a how-to-do-it handbook you inherited from the old guy down the hall.

We could call it "irreverent reality," a term I coined to describe my colleague, Karen Eitreim, who had a passion for kids, schools, and serious scholarship, but also knew how to laugh—after her morning latte.

Foreword

Don't let the "aw shucks style" fool you

When Jim asked me if I would "say a few words" about his book, *Taking the Pain out of Pre-Algebra*, I thought he was joking. I was an abysmal math student in high school. I was a "no math student" in college, and because of that, I chose to major in Secondary Education, rather than my preferred Elementary Education major. I wanted to avoid taking two required math classes. Little did I know that as a graduate student several times over, I would regret that opportunity to sharpen my math skills and overcome my math anxiety.

Over the last year, I have followed the evolution of this pithy handbook and Jim's enthusiastic work as a long-time math teacher in several area schools. His experience spans parts of six decades. Jim's passion for his work, his students, and his approach to raising the bar for struggling students has been inspiring to me. As a career teacher and a school board chairman, I am intimately aware of the decline in math skills of middle and high school students.

What I found when I reviewed this book is that it is NOT a book about the "science" of teaching Pre-algebra. Don't let Jim's casual register, the cartoons, and side stories fool you, this is a down-to-earth, no-frills guide to the craft of teaching that can be applied by teachers willing to make changes in any upper primary, middle school, high school classroom. It addresses behavior strategies for effective classroom management, organizing instructional materials, and scheduling a week and a year of work while establishing a system of accountability for students, parents, and administrators. Threaded throughout is the emphasis on the social-emotional needs of struggling students that can be addressed by deepening expectations and giving the students the structure and support they need to believe they can master Pre-algebra.

Jim has shunned the scholarly approach. He only hints at the landmark pioneers of instructional improvement, yet their work and advice is clearly woven throughout the whole book. William Glasser would approve of the ideas for building classroom community. Every lesson resonates with Madeline Hunter's ITIP (Instructional Theory Into Practice). His grading strategies echo Thomas Guskey. If you use Jim's methods, you should fare well in a more modern Charlotte Danielson evaluation model.

I recommend this book to teachers newly assigned to Pre-algebra classes. I also recommend it to student teachers and to new teachers regardless of the content area. It is a resource for Title I Interventionists, and special education teachers, and an excellent "book study" for PLCs in middle schools, and as Jim would say, "The stories are pretty good too."

Susan Watts Day, PhD

Special Education Teacher, Retired
Elementary School Counselor, Retired
Teacher Education Associate Professor, Retired

Chapter 1

What do they call it at your school? Pre-Algebra?
Yeah, I know. You didn't want to teach this class.
Who Does? Pretty much nobody.

Most likely you're a beginning math teacher paying your dues until you can get some "real" math classes. You could be from a more enlightened school where everybody takes a turn in the "low-end" math classes—rare, but it happens.

Or, ugh, **they** might be filling up your schedule because enrollment is low in your regular classes. Or, worst of all, **they** assigned these kids to you hoping you'd quit.

Now you have a choice. You can trudge through this class every day with one eye on the clock while you endure the soul-crushing apathy of these kids and their lousy social skills.

–OR—

You can do something meaningful and help these kids improve their math skills, their social skills, and their learning skills. At the same time, you can improve your instructional and classroom management techniques. When you get good at teaching math to kids that struggle, you might even start to like the class.

There is an upside to this assignment. **They** pay no attention to this class or these kids. Nobody really cares what you do with the curriculum and instruction. Keep the office phone from ringing, don't write up too many kids, and you can pretty much do as you please. It's not much of a kingdom, but it's all yours and you can turn it into a reasonably pleasant experience for yourself and the kids.

I'm not a real math teacher—I was a shop teacher, an alternative school principal, and finally a math teacher working exclusively with struggling students.

I am the only teacher, that I know of, who specifically asked to teach these difficult math students. I'm good at it. I spent eleven years learning how to make these knuckleheads more successful—much more successful than they have ever been since second or third grade. I can show you how to do it too. So follow along as I explain Success Strategies for Struggling Math Students. The worst that could happen is that things won't get any worse. Things might even get better. Let's crack a beer (wine is okay too), and figure this out.

So why bother with this book? Why would I take a couple of years to put in the life-stealing time to write a how-to math book? I am seventy-five years old and beginning my fiftieth year of teaching. I don't have a lot of time left, and somebody needs to document how things could be better for these kids and teachers in these low-level classes. Struggling students, and even capable students, can enjoy much better outcomes when you employ success strategies. Once you learn these techniques, you will get more done in less time with less effort.

Essence of this book: You can't buy better math scores; you must earn higher scores with better instruction that fits these kids.

I'm the old guy on the left. The other old guy is Tony, my teaching partner in Yelm, who helped to develop some of the Success Strategies. Even when we're sailing, we talk math. I wish you could join us. We would share our passion for these kids.

There is no way to improve math success for struggling students unless teachers are willing to change the way they teach. There is no known correlation between improving math learning and a host of, instructional practices, such as:

- Putting assignments on Canvas or Google Classroom.
- Writing a three-digit code for target learning/goals on the board.
- Using graphing calculators.
- Buying new math programs or better books.
- Using "smart boards."
- Refusing to accept late work.
- Lowering grades on make-up tests.
- And add whatever you want to the list

And by the way, here are some more things that don't correlate with math success: master's degrees, board certification, and long years of experience. Some studies show that non-math majors don't do as well as real math teachers. If you buy me a beer, I'll explain what that research is bogus too. For some nachos, I will also explain why the class size studies are deeply flawed.

Small, but statistically significant, improvement in learning occurs when the curriculum is aligned to the test. Makes sense! Kids are likely to get higher scores if you teach them what is on the test, but it's only a one-time fix, the low-hanging fruit if you will.

Math videos on YouTube could help your students, but the struggling kids we're talking about do not watch Kahn Academy or Brian McLogan. They didn't watch in the past and will not be watching in the future—you are their resource, their only resource.

Chapter 2

So, Jim, you think you're smarter than we are?

Definitely not smarter, but I caught some breaks you never had.

- I wasn't burdened by a lot of math success in high school.
- Our high school math teacher only knew how to work with kids that "get it" quickly—and coach basketball.
- I redid all my high school math in the army. (Algebra I, Algebra II, and Geometry) so I could pass Calculus in college.
- I was the worst beginning teacher I ever met. Barely passed student teaching. So I set out to become the best teacher I ever met, and I worked at improving every single period of every day for almost fifty years.
- I taught shop and learned that every kid needs to learn everything—every day, lest they go home without thumbs.
- I learned how to teach the whole group and still connect with kids one-on-one.
- I found my passion for working with difficult kids (knuckleheads).
- I was the principal of a hardcore alternative school. I honed my discipline philosophy and skills. It was probably the worst high school in the State of Washington, and we fixed it.
- I have been teaching difficult, struggling kids for longer than most of you have been alive.
- I learned to spell efficacious and three other hard words.

John Hattie is right, "Teacher Efficacy" is the single most important variable in learning. In this little book, I will show you how to achieve greater efficacy for the first three stanines (the lower achieving third) of math students. Most of you will resist, just as my math-teaching friends resisted.

I don't think this book is a waste of your time because I (we) proved it works. First, it worked in my classroom at Elton High School. Then it worked in three other classrooms in three other schools with three different teachers.

efficacious adjective

ef-fi-ca-cious

definition of efficacious

:Having the power to produce a desired result

//an efficacious method

I can also spell mayonnaise without using auto correct.

Our little group in Elton was so successful, the boss hired some "regular" math teachers who were supposed to teach using my methods and materials in "regular math kind of way." It did not work. The real teachers just couldn't get over the "stand-and-talk" method of teaching.

That pattern followed everywhere as fifty-eight school districts and about 150 schools implemented "success strategies." Mostly math teachers could not (would not) make these ideas work. BUT given the same in-service and the same materials, shop teachers, special ed teachers, science teachers, and a few outlier math teachers got similar results to those we enjoyed in Elton—two to three grade levels of improvement in one year. If you can do that for two years, struggling kids almost catch up to grade level.

Turns out "success strategies" works for high-achieving kids too. After I retired for the fourth time, I was hired to create a class for students—seniors and juniors—who had not passed the state math test SBA (Smarter Balanced Assessment) nor had they passed the ACT. The success math methods worked there too. In my school, 100 percent of 120 the students we worked with (there were two teachers) passed. They had similar results at the other high school that used our methods. Almost nobody passed at the third high school where the teacher ignored these ideas. She kept lecturing, the kids kept screwing around, and she scolded and threatened. She cried, and they jeered. Nobody passed.

Later, I also applied many of these methods in a conventional Geometry class. It worked there too. And—even later, I took over an Algebra II class with some pretty darn capable kids. It worked there too.

This book challenges almost every idea about teaching math in a traditional way. It is, at the same time, both revolutionary and simplistic. As you already know or suspect, there might be better ways to teach low-achieving students. But you never had the time or support to put it all together as a comprehensive approach to teaching kids who hate math.

The first year I tried Success Math, I stumbled. Sometimes I would rewrite a lesson after the first period and teach it again in the second period—better results. Now and then, I would rewrite it during lunch and teach it in a slightly different way during the third period. Always, at the end of the day, I would rewrite the lesson again for use next year. I also added notes about how to work with the students during instruction. This cycle went on for eight years. This book is not about genius or some slick, easy idea. It is about grinding through 4,320 tiny iterations of improvement until I had something that helps struggling students learn math.

Most math teachers just wouldn't do it. Some like Tina just sat with their arms folded (more about her later). Some were openly hostile to these ideas. Some were more honest. My friend Don said, "Jim, I know the method works, but it takes too much time. I would have to learn new ways. I like my coaching and hobbies and time with my family. I already have all my slides made, I am going to stick with my overhead."

Another colleague, Nancy, an excellent math teacher said, "Jim, I know this works, but I became a math teacher because I like my nice, polite, kids, sitting in rows, quietly scribbling away at their math work."

Sometimes it's like trying to get an elementary teacher to drop her dinosaur unit.

My less kind colleagues suspected me of cheating when my kids' scores increased dramatically. Those two teachers backed down when they heard the non-rumor that I had retained a lawyer.

In this book think of "I," "we," and "you" as equivalent terms; we are doing this together. ***They*** *mean the dark forces of the district office and administrative people that want you to stick to the standard script even though it does not work for struggling kids and never will work.*

I still substitute teach a couple of days a week. I prefer the more difficult kids in lower-level classes since pre-calc kids don't like me interrupting their work with my stories. I hope it keeps this book timely. I am still in this with you.

Most districts have a long, arduous policy to approve instructional materials. But there is usually an exemption for "supplemental" materials. If you stay quiet and get good results, ***they*** *will probably never ask how you did it.*

Chapter 3

You need an assessment system and here's a trick or two.

You need an assessment system. Your first goal is to find the kids that were misplaced in your class. Every year, I identified a few students who were incorrectly stuck in my class based solely on their poor grades or one bad test score. You do not want these kids in your class. They are ready for Algebra I and they are going to be angry and upset. As the semester progresses, they will make you angry and upset. Identify these few kids the first week and get them into an appropriate class. Everybody will be happier except the counselor who must make the schedule change.

Your appropriately placed math students will be dramatically improving their math scores. You need to document this improvement. The more successful you are, the more traditional math teachers will question your scores.

The best bet is a test of some sort that provides grade level equivalents. People don't understand arbitrary scales such as the SBA or MAP. They really get 2.5-grade levels of improvement The curriculum office may resist your efforts to convert to grade levels, but if you press your case, they will help you. If they continue to balk, read the district mission statement to them. They won't like it, but they will find something just to shut you up. If they won't help, you can find tables that translate arbitrary scores into grade equivalents. Hard to find, but they are out there.

Here's the trick—sort of: Give the kids the first assessment very early in the school year while they are still packing their nasty attitudes about math. They won't try very hard. Most of the students will act all surly and give up before they even look at all the questions. That's okay! It's an honest snapshot of their feelings and willingness to work and learn. That's their normal, and you deserve credit for fixing math skills as well as math attitudes.

Typical Pre-Algebra student during state test.

Later in the year, because they trust you, the kids will try harder. Just because they like you, they will make a real effort. And because you taught your students the math they missed along the way, the scores will increase even more. Don't over-test. Do the first assessment early in the year and the second one late in the year. In between, just stick to your unit tests, that's enough.

Two more tricks: Come up with a big reward for honest effort on the state test, and another reward for trying hard on your final class test. (Summative assessment if you want to sound all "Mathy") My partner, Tony, and I promised the kids that every one of them who made an honest effort on the state test would be invited to our barbeque. Hot dogs, buns, chips, potato salad along with store-brand soft drinks didn't cost much. We called in sick on test day so we could be available at the test site with our kids. We met them at the door with sharp pencils, a pep talk, and some candy. During the test, we walked around the gym watching the students work. If they were working hard, we laid a ticket to our barbeque on the table. Kids that were dogging it saw what was happening and picked up their

A student will do a surprising amount of work for just a hot dog and some chips.

efforts. In the end, only five out of ninety kids didn't earn a spot at the barbeque. You will need some guts here. We held tough on the five, and they didn't get to attend. Word got around. The following year, every kid came to our special luncheon.

Regular math teachers cried foul. "You guys bought those higher test scores." Our answer was, "Okay, we bought higher test scores. It cost a whole lot less than our last Marzano workshop." We didn't even tell them about the ice cream sundaes we served as a reward for good effort on our in-class assessments. A few gallons of ice cream and some toppings don't cost much at all.

We did have one parent call, "Why was my kid excluded?" Luckily, we had a great principal. He told her, "Danny didn't work during the test and everything about this was done on their own time with their own money. I couldn't order them to include Danny even if I wanted to—which I don't." (Danny was a pretty much an infamous jerk at school.) Then mom called the Supe who pretty much told her the same thing. Sometimes, you need a sponsor or two when you want to stretch the boundaries of "normal". I cover this idea in some detail in book two: *How to Fix a Broken School.*

I have simplified the MAP and SBA conversions charts for you. Use these charts with caution. They are old, and they have been summarized. Use current information to make academic placement decisions and communicate with parents and students.

Smarter Balanced Assessment Scale Score Ranges for Grades 3-8

Level	Grade 3	Grade 4	Grade 5	Grade 6	Grade 7	Grade 8
Level 4	2501-2621	2549-2659	2579-2700	2610-2748	2635-2778	2653-2802
Level 3	2436-2500	2485-2548	2528-2578	2552-2609	2567-2634	2686-2652
Level 2	2381-2435	2411-2434	2528-25-78	2473-2551	2484-2566	2504-2585
Level 1	2189-2380	2219-2454	2455-25-27	2235-2472	2250-2483	2265-2503

MAP Score Translation Chart Sample Scores

Illustrative only

Do not use this chart to make academic decisions or to communicate with parents.

Grade Level	Fall	Winter	Spring
4	198-208	202-213	206-217
5	206-217	210-222	213-226
6	212-225	215-228	218-231
7	218-231	220-234	222-236
8	223-237	225-239	226-241
9	225-240	226-241	228-242
10	229-244	230-244	231-246

Chapter 4

Who are these guys? Why do they struggle so? Why won't they work? Why are they so lazy? Why are they such rude jerks?

"I'm a good math teacher. If I had some decent kids, I could . . ."

There is no sense complaining about these kids. Their scores place them in the first three stanines of math achievement. Their parents send you the best kids they have. They aren't keeping the "good ones" at home. Most of these kids have hated math since elementary school. They are what they are, and they are not going to change. They are not like you. If you keep teaching them the way you were taught, you will never get anywhere. Teaching them the way you were taught will not work any better this year than it worked last year. It won't work any better next year either.

If you want to be successful, **YOU MUST CHANGE THE WAY YOU TEACH. And absolutely, you need to change your attitude about these kids.**

Mostly all of us believe that other people are like us. Which is kind of a natural conclusion since most of us have never been anybody else. But with some patient observation and reading, maybe we can learn to see the world a bit more the way other folks see it.

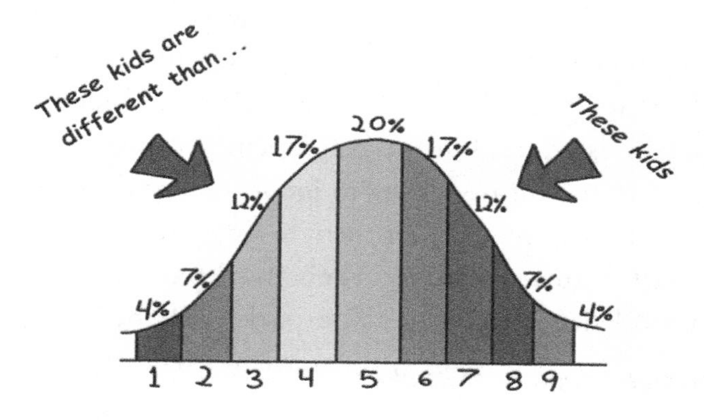

It okay to teach struggling kid in a different way than successful kids. They are different kinds of students. They are different in ways we can know and measure.

Some things you need to know about struggling students and math teachers.

Certainly, the traits we'll discuss are generalizations, but they are generally true most of the time for most people. (I have tested a thousand or so struggling math students and hundreds of math teachers to reach the following conclusions. It would be helpful if you would take a Myers-Briggs survey or attend a class to verify these assertions.)

There are dozens—perhaps hundreds of knock-off MBTI indicators available. If you complete the MBTI with a trained teacher, you may gain insight into yourself as well as your students learning style.

P vs. J: Mostly the kids are *P* (perceiver) learning style. Most teachers are *J* (judger) learning style. Struggling students procrastinate, it's who they are. They delay, postpone, forget, and don't worry too much about a missed deadline or assignment. A consequence of more than an hour in the future is easy to ignore. A consequence a month away from today is impossible for them to imagine.

You shall know them by their backpacks. Look at your struggling students' backpacks. The packs will be stuffed full of wrinkled, dirty, folded, half-finished assignments that never got completed. Some of the assignments will be done, but even the finished ones never got turned in. In a later chapter, I will recommend that assignments are never allowed out of the room.

Recognize these backpacks?

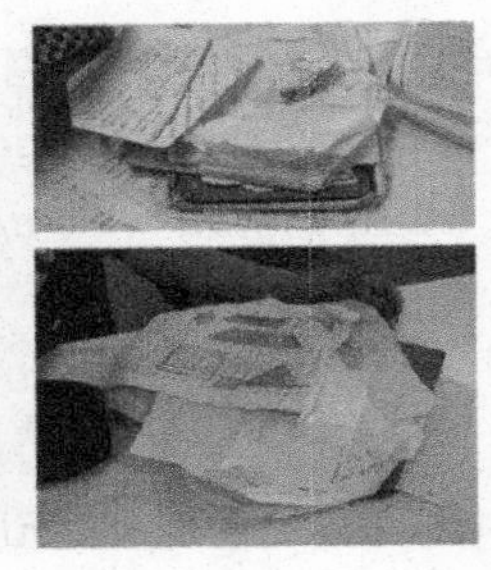

Real, live backpacks from my Pre-Algebra class. This has not changed in the 20 years. I have been keeping track.

Struggling students are very apt to show up at your class without a book, pencil, calculator, or paper. If they bring a computer, it's likely not charged. They have never made a calendar or a "to-do" list in their lives, and they won't be making any such lists or calendars in the future. They struggle in school and math precisely because schools most value the things that the students most lack: : punctuality, preparedness, and completeness.

Most math teachers are *J* personality types. Teachers thrive with deadlines, planning, and punctuality. (Extreme *J* types live for the check marks on their lists. If they complete a task that wasn't on the list, they will add it to the list just so they can check it off.)

These traits helped them become good high school and college students. Teachers tend to believe that *P* learning type people are lazy goof-offs and that it is the teacher's job to help them learn these *J* skills through a series of punishments that will help them become responsible. Teachers believe that *J*-ness can be taught. Teachers will have a lot more success if they learn to work with student *P*-ness; it can even be a positive thing for learning math. We will cover this in some depth in the chapters on direct instruction and daily assignments.

I can promise you that you will not be changing the basic personality trait of procrastination. *F* grades don't bother these kids too much, and penalties for late work certainly don't motivate them even a little bit. If you have already received six or seven *F* grades, another one is no big deal.

F vs. T: Feeler types (F) and Thinker types (T) are generally a 50:50 distribution in the population. Feelers value relationships, friendship, loyalty, and socializing with friends. Think mercy/kindness. Chit chat first, and work second if there is time. Thinkers value rules, predictability, accountability, and consequences. Think justice/fairness. They like to work first and visit later.

My wife, Pat, max on the feeler scale, says it is better to be a thinker (like her husband—max on the thinker scale) because it's much easier to think about feelings than it is to feel about thinking. I could fire a friend if needed—I wouldn't like it, but I could do it. Pat would probably quit her job before she fired a friend.

I bet you can guess where this is going. Somewhere around 75–80 percent of struggling math students sit on the feeler side of the Myers-Briggs scale. About 75– 80 percent of math teachers fall on the thinker side of the MB Type Indicator. Here's the conflict. Kids want mercy. They want teachers to understand why they can't get their work done. They want teachers to make allowances for all the things kids couldn't control about their lives, and for many of them, there is a great deal they can't control.

Teachers want kids to focus on class work instead of answering the "urgent/important" text from a friend in crisis who is breaking up with her partner. Teachers want stuff turned in on time, sorted into boxes, with legible names at the top right corner. Teachers want results instead of excuses.

E vs. I: This Myers-Briggs type is not such a wide gap as some of the others. About 75 percent of the population are, in varying degrees on the extrovert side of the scale. About 25 percent of any random group are likely to lie on the introvert side of the scale. (Except of course for engineers who are almost exclusively introverts.)

Math Lab MBTI Scores (2004—2005)

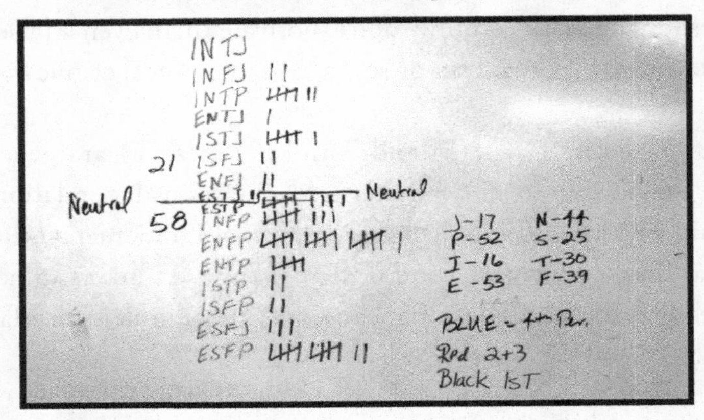

This group of ninth graders averaged three years of growth in computational skills. None of them passed the 7th grade state test. The scores for math teachers are almost a perfect inverse. Did you notice--not a single INTJ in the bunch.

As you might guess, math teachers tend to be more introverted than their students—especially their struggling students. In the groups I work with, about half the math teachers register as mild to extreme introverts. That doesn't mean that math types seek the solitude of a hermit.

It does mean that math teachers are fine with working alone for long periods of time. They appreciate quietness while they are concentrating on learning a new skill or making sure all the details are correct. They tend to listen before they talk; they test a sentence in their minds before they blurt it out.

Extroverts? Not so much! In a class for struggling students, you can safely assume that 80 percent of the students by varying degrees are extroverts. Extreme extroverts start speaking before the sentence is fully formed in their minds.

Whatever pops into the head of an extrovert is likely to come out as unfiltered speech. It doesn't matter if the time is right or distinctly wrong. It is almost impossible for extroverts to let a thought go unsaid. They relish working in groups or at least pairs, they abhor aloneness and quiet.

This is the students' style. It is not your style. You cannot make enough rules or sanctions to change the way they interact with other people. If you want to be the teacher that gets these kids to succeed, you will have to convert your class so that their extroverted social needs turn into math-class positives.

Math Intelligence. Math makes sense to math teachers; they possess a high level of what Howard Gardner calls logical/math intelligence. Struggling kids have much less native ability. They don't easily recognize patterns and repetitions. Input-output systems are not obvious—especially if they are called functions with the notation "f(x)" instead of plain old "y." Struggling students are more likely to see the world in a qualitative way than from a quantitative perspective. How does it feel and taste? Does it make me happy? The "**What**" of things holds more importance than the **"How Much"** of things.

Lacking this "knack for numbers" and "if: then thinking" math is difficult for them to learn and even more difficult to remember. **Forget your pacing guide.** Learning a new skill will require many more iterations of practice than your textbook anticipates.

Just for fun, go check out the eighth-grade standards for math. You will find three and a half pages of small type. Look carefully and you will wonder, "Did these people ever actually work with a struggling math student?"

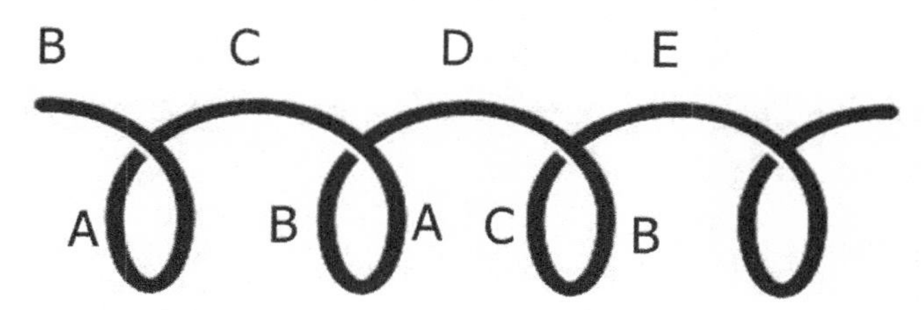

Rolling reviews means that we keep looking backwards now and then as we move through the semester and the year. Even when we get to goals W and X, we should look back at goals A and B now and then.

More importantly, your struggling students don't retain new learning for as long as teachers do (or did when they were young). Each newly acquired skill will need frequent review over the course of the entire school year. I call it **rolling review**; I stole the idea from the Saxon math books, but the words are mine.

Calculating: Math teachers don't have to think very hard to know that 5,026 x 3 is about 15,000 and 75 or so more. In just an extra second or two, you can validate that approximation with an answer of 15,078. You will arrive at that answer long before the students can retrieve their calculators from their backpacks or fish out their phones. In daily teaching, you will have moved on long before the kids turn the calculator on and key in the numbers and operations. Struggling math students are profoundly handicapped in the skill of mental math and simple calculations. They do not—never have—and probably never will know their math facts.

These kids do not know the product of 7 x 8, and .7 x .8 is a bigger mystery than the pyramids. I discovered that you can with near 100 percent accuracy predict who is guaranteed to fail standardized math tests based on their ability to do 25 multiplication facts in one minute. See Appendix C for a sample test you can use. If kids cannot multiply, they cannot divide. They can't make estimates. They cannot test the reasonableness of the number that magically appears on the screen of their calculator. Their skills are so low that most of them will need a calculator to come up with 20 when they see 17 + 3. It may be fashionable to write off this skill deficit as "calculators are everywhere—understanding processes is what really matters." Even if that were true—and I don't think it is—you, as the teacher, must work with kids who don't know or see simple calculations as easily as you do.

The sad part of this picture is that it's real. This youngster could not find his calculator, so he grabbed a white board.

I never met a math teacher who lacked command of math facts. There is a work-around for this calculator dependency that does not involve doing weird finger counting on knees hidden under desks, but you need to be mindful of the students' disabilities as you plan lessons and assignments. You can teach these students algebra more easily if you stick to easy numbers. You need to utilize numbers such as 2, 3, 4, 5, 6, 8, 9,10, 12, 20, and answers that are whole numbers. Don't let the lack of basic skills obstruct learning the steps of algebra. You will be amazed at the difficulty of creating math problems with nice even, whole number answers.

Your kids never really learned their multiplication facts, and they will resist your efforts to help them.

Sensitivity to criticism: Struggling math students possess a keen awareness of their difficulties. They believe that you are born with math ability, or you are not. So does most of American society. Students translate their lack of ability into one word, "**dumb**." It undergirds all of their expectations and behaviors. They go to extreme lengths to avoid "dumbness." Better to appear lazy than dumb. Better to act out in class as a clown than seem "stupid." Better to withdraw under a hoodie with sullen silence than to speak and reveal "dumbness." Their words, not mine.

To help struggling kids, you must not confirm their fear of dumbness. It is not helpful to point out that, "You should already know this." Equally unhelpful is the question, "Who was your teacher last year?" or "Where did you go to school last?" You must avoid even the slightest hint of sarcasm or disdain. No eye rolling! No sighing! No looking at the ceiling! No deep breaths accompanied by chest rising! No head shaking! And most of all, no snappy, sarcastic comebacks to their lame excuses. These kids may be lousy at math, but they're people smart and acutely aware of subtle expressions in human interactions. You cannot shame or embarrass your struggling students into more effort.

Yes! Sometimes discipline is an issue. You could certainly argue that the kids have learning problems because of their classroom behavior—OR, if we dig a little deeper, we might discover that much of their classroom behavior is driven by their past failures in math.

Let's admit it, sometimes these kids are extra special difficult to deal with. Discipline issues are the main reason most math teachers (or any teachers) don't want to deal with these kids. They like to be off task. They talk too much—especially when you want to talk. They are often disrespectful—sometimes they are downright confrontational, and they appear to be lazy—ALL TRUE. That's why there is a whole chapter devoted just to discipline. But here is a big hint. Many of the discipline issues are co-mingled with fear of failure, fear of "dumbness," past bad experiences, and life in general. In many ways, these are damaged kids, and we school people have contributed to that damage. And yes, some of them are just little shits! But almost all of them can fix themselves with help from us.

So how to deal with the fear of dumbness? It kind of depends on your theatrical ability. My personal schtick goes something like this. "I am the only math teacher you will ever meet who hates math. I wasn't very good at it, and I had crappy math teachers in high school. But in the army, I decided to redo all of my high school math again. I learned Algebra I in a month. Algebra II took six months. Trigonometry took fifteen days. I never did pass all three quarters of college calculus, and I hate geometric proofs—but I can do them.

It's my schtick, and I'm sticking to it.

I am kind of like your math angel. I was sent here to help you learn. **Math is not hard. Math is boring. Math is tedious. Math is not especially fun.** It is just a bunch of procedures you need to memorize. It has a whole lot of very picky vocabulary. But math is a heck of a lot easier than learning how to use your cell phone and all that social media stuff like tick-chat or snap-gram."

Chapter 5

First things first.
If they don't believe you like them, they won't do diddly.
And you can't fake it.

Just barely fifteen, Bobbi turned to me, "Slosson, do you really like us, or do you just pretend to like us so we will get higher test scores?"

"Well, Bobbi, that's a pretty smart question. It takes a pretty smart person to ask that question, and you are a pretty darn smart person. Would you agree?"

"Well, I'm not school smart or math smart, otherwise, I wouldn't be in your loser math class. But yeah, I'm people smart. I guess I am probably life smart, I mean as smart as you can be when you're just a kid."

"Okay," I answered. "Do you think I could fool a smart person like you for four and half months? Wouldn't you figure it out if I were faking?"

Her face lit up, "No, you couldn't fool me for half a year, nobody could fool me that long. OH MY GOD! You really do like us. How could you possibly like us? We're rude and disrespectful; we talk to each other, and we don't listen to you when you try to tell us stuff. We're lazy, we don't do any work. We screw around in class. We don't do any homework. We skip all the time. Everybody hates us—especially math teachers. Face it, everybody hates this loser class. Why don't you hate this class? What's wrong with you anyway?"

"Hmm," I squinted over the top of my glasses. If you make an honest effort on the state test—you don't have to pass—I will tell you why I think I like you guys."

I must share with you that Bobbi was way more sophisticated than her fifteen years. The first or second week of class, she told me, "Everybody hates you. You know that?"

I had a pat answer left over from my days as an alternative school principal, "Yeah, Bobbi, I get that a lot. Most kids don't like me at first, but they come around. I'm a nice guy. I am not your friend, but I am your friendly math teacher. Only a deeply flawed person could dislike me for a whole year."

Bobbi turned to her buddy, a thug named Lance, and kind of poked him, "Hey, Lance, you'll probably hate him for two years."

Bobbi didn't pass her state math test, she wasn't in school by the time the test rolled around. There were lots of problems. She dropped out.

Got an AA degree and some specialized training. At age twenty-nine, she had kids, a solid marriage, and a really good job managing the office and support staff in a large accounting firm. She likes the work, she excels. Bobbi likes using her skills to work with younger staff.

Okay. One more story before we will get to how you can let your kids know that you like them more than you like math.

I was teaching an in-service for a school district in NW Washington. I had given them my curriculum and instructional materials. But it was a gift with strings attached. As part of the license agreement, only teachers who sat in my two-day in-service were allowed to use my stuff. And yes, I did charge them for the in-service. And yes, some folks didn't want to hear what I had to say.

I was surprised—almost shocked—to see Tina sitting in the front row of this second in-service. I did the normal intro, "Can anybody here tell me the quadratic formula without looking in a book?" Every hand went up. Unless I have some very honest shop teachers or special ed teachers in the room, that is what always happens. I go on, "Great! You all know more math than I do. Now that we have established that, you don't need to spend any time at all trying to prove you're smarter than me. You are smarter."

I turned to Tina, "Tina, I am really surprised to see you here today. The last time I was here, you only attended because you were forced to be here. You sat in the back row with your arms tightly folded, staring at the clock for two days. You didn't like a thing I said. If somebody is making you attend today, you can go. I will sign your attendance sheet. I am not going to make anybody suffer through this twice."

Tina was not receptive to the emotional intelligence part of our in-service. And then...

Tina looked a little sheepish and said, "I'm sorry, Jim. You're right, I wasn't very nice." Then she stood up and faced the rest of the class. "Last semester, I had the worst class I have ever had. They hated me. I hated them. My stomach would start aching the hour before they came into my room. The clock crawled, somebody would always have to watch my next class for a few minutes so I could go to the bathroom."

Tina continued, "I thought to myself, I have nothing to lose here. I'll try the interviews. We did it just the way Jim will show you in this class. I couldn't believe it! Before we were halfway through all their interviews, the kids started liking me, and I started liking them. By the time we finished the interviews, I was beloved. The kids started to like my class. They even learned some math—not all that much, but a lot more than they knew before. Most importantly, they were willing to be polite and willing to make an effort in math."

Then Tina turned to me. "Jim, I really am sorry about the last session, and this time, I will pay attention. It works."

And the moral of the story is: If you like them, they will like you. If they like you, they will be much more inclined to do some work for you. The worst that could happen is that your class will not be a little slice of hell every day.

I should share with you, however, that this is a powerful technique. I once used it to take control of a jury. Then I felt so bad about my manipulation that I turned them down when they wanted to elect me to be the presiding juror. Then I had to re-assert control when the jury went off the rails. But I think you will have to buy me another beer to hear that story.

Tina started with the student interviews because it is the most powerful way you can let the kids know that you like them. The interview works because it shifts the focus of the class from tedious, boring numbers to the kids' favorite subjects, themselves. It works because it gives you the opportunity to learn a little more about each student. The interview helps students establish a sense of connection with other kids in the class. It works best if you can have the students sitting in a circle while they do the interviews, but it can work in any seating configuration.

A set of interview cards is included in the Math Lab/AGA materials. You can find the instructions in the Appendix section.

How the interview works: One student is the interviewee and sits at the head of the class. The rest of the class asks questions in turn. To make this work, you will need some structure and rules, or it will turn into chaos—thus the cards. The cards are printed with the questions and distributed to the class. The monitor—not the teacher—calls the card numbers in order. The interviewer (who is every student in the class) says the student's first name and then reads the question. The interviewee answers the question. The interviewer may ask one relevant follow-up question. The monitor asks for the next number for the next interviewer and so on until all the questions have been asked.

A major part of the interview process is that all the students will begin to learn the names of all the other students in the class as well as a few things about the student.

The questions are safe and simple things such as: What is your full name? Which part of the district do you live in? Do you have regularly assigned chores? How many siblings? Favorite fast-food place? And so on. Students do not have to participate. BUT in fourteen years of doing this, I have only had one student who refused to be interviewed. The rules are you must be respectful and listen—your turn is coming. Kids love this.

Tip: If your supervisor comes in to observe and your lesson is a little lame, let the kids interview the observer. Warning! Adults, especially administrators, can pretty much use up a whole class period. They enjoy a pleasant non-confrontational, conversation about themselves. Their answers tend to be long.

Ordinarily, with one question per student and some practice, you can get an interview done in about ten to thirteen minutes. The number caller of the day (moderator) can pass out the cards as students enter the room. After a few iterations of the process, kids will already have some ideas for their responses so they will answer quickly. And the activity gets students settled down and ready to focus on what you would like them to learn for the day.

Once every student has had an opportunity to be interviewed, you will be tempted to drop the starter activity at the beginning of class. Resist that temptation. Continue to do some starter activity almost every day. You can do times-table races, you can play the anticipation game, you can move on to other games and activities, and you can do more interview questions. You can do the book of questions. You can even do old-fashioned elementary school show-and-tell. There is a long list of games and instructions in the AGA-Math Lab curriculum.

This team-building time is not wasted time, it makes your instructional time far more effective. But you must summon up some courage to use 15 percent of your class time on social-emotional growth.

Your colleagues will not be supportive. In their view, you are wasting time and coddling kids. Once you begin to be successful, and students enjoy coming to class, the other teachers will be resentful. Gadzooks! Someone may ask them to humanize their classrooms too.

If you are really pressed for time, you can change this "Interview" to "Hot Seat." If the lesson went well, and most of the kids seem to be done near the end of the period, select one student volunteer for the interview. The student sits in front of the class, and you rapidly ask the questions—whichever ones you think are suitable. This goes quickly, you can ask twenty questions in less than five minutes. After a few sessions, the kids will want to read the questions, and you can appoint one as the reader for the day.

More team building—Who's Liz?

I was the long-term sub, taking over the geometry class in November. I only met the class four days before, they sat in a classic row formation. The students never experienced a math teacher walking around checking and correcting their non-homework assignments. The backpacks dang near killed me at first.

I asked one student to help another. Both the students were surprised when I asked Jacob to let Liz help him with a factoring problem. Jacob asked, "Who's Liz?"

I answered, "Are you teasing me?"

Jacob said, "Huh?"

"Jacob, really, you don't know the name of the person who has been sitting next to you since September? That's almost three months. Liz is sitting to your right."

Jake looked at me like I just dropped out of the sky, "No, why would I know her?"

I looked at Liz, "Do you know his name?"

"Well, Mr. S, now that you've said it, I guess he's Jacob."

"Well Jacob, may I present Liz, Liz, this is Jacob. Liz, would you be so kind as to help Jacob with the 'think-and-dink' part of this factoring?"

Why would I know her name? We're not allowed to talk.

Big smile from Liz as she turned toward Jacob. "Nice to meet you, Jacob, would you like some help? I can explain it better than he can (meaning me, the teacher)."

I had to ask. "Seriously, you guys don't know each other's names?"

Jake said it best, "Why would we? We never talk. We're not supposed to talk in class."

"So in this class, you guys don't help each other?"

Liz looked up, almost offended, "We're not cheaters!"

Here is an interesting experiment you can try in your class. Hand out a blank seating chart to each student. Ask the students to write their own names in the correct spot on the chart. Then ask for total silence. "This is a little non-graded test, but I may have a treat (I use Hershey's Bars) for anyone who can name all of the people sitting one space around them." You will be shocked. Even if the kids have been sitting in the same spot in the same row for most of a year, you will discover that almost nobody knows the names of the four people who sit adjacent to them. There will probably be one extrovert who can fill in the whole chart, let that person take roll when you have a sub.

What a shame! How can these poor kids be so disconnected from each another? What a drag it must be to go to class every day, sitting next to strangers and trying to kill an hour not listening to some teacher who goes on and on.

Were things different, the people nearby could be teammates, acquaintances, mentors, maybe even friends. And you, the teacher? You could have a classroom full of student-tutors helping each other. Once you got the hang of it, you could wander around the room and see who needs what kind of help. You could ask kids to help each other, they might be willing to help and accept help because they are connected at least by their names.

If you start this naming process with the interviews, you can use simple activities to expand it. Go ahead and take roll orally—yeah, I know the "teach-bell-to-bell" crowd doesn't like that, but it's your class. You can do something like, "Trevon, is April here today?" (I like to start the semester using an old-fashioned attendance sheet, sometimes I stick with it for the whole year) Early on, I just go down the list alphabetically. I asked Trevon Anderson if Victoria Baxter is here today? Each student only needs to learn one name.

Tre responds, "I have no clue who April is."

"Well, look around Tre, do you see a glimmer of recognition? A smile? A shy turning away?"

Tre looks at me, "Is it her?"

"Don't ask me, ask her," says I.

Then it's Victoria's turn. "Victoria, is Tanner Collins here today?" And so on.

It's a bit brutal at first, but they catch on quickly; they are social animals.

After the students know the name of the person that comes after their name, you can do it backward. Work from the bottom of the list up, ending with, "Victoria is Trevon here today?" After a time, you can ask the first student on your roll sheet if the third student is present. "Trevon, is Tanner Collins here today?"

Kids are more likely to want to come to class when they are not strangers. When all the kids know the names of all the other kids, they will work more cooperatively, they will learn more. Your job will be easier. You will have twenty-three classroom tutors.

Cheers had it right. "Sometimes, you want to go where everybody knows your name, and they're always glad you came."

Chapter 6

Rows Suck—Circles are Best—Table Groups Work Pretty Well
What works in a tavern will work in a classroom too.

In Washington State, we used to have a law that required all parts of a tavern to be visible from any part of the tavern. The room had to be a box, circle, or oval—no "L" shapes allowed. No matter where you were standing, you could see everything. The same idea works splendidly in a classroom full of struggling math students.

Bob said, "Slosson, I hate these tables. I hate the circle. I like to sit in the corner. I don't like you looking at us all the time. Plus! you never turn your back. I mean damn! Man, you don't even let us sharpen our pencils."

"Well, Bob, you like to sit in the back corner so you can mess around and not do your work. And you don't need a pencil because there are always plenty of sharp ones when you walk in the door. And I have the room laid out like a tavern so I can see all of you all the time even when I am working with one or two students. I got the idea in a class I took." (TESA: Teacher Expectations-Student Achievement). "Most teachers only talk to the kids in the front row and the middle row. Guys like you prefer the corners." I handed him a Hersey's Kiss. "Stop whining. It's unbecoming a future welder." I moved on. He went back to work.

And the moral of the story is: Kids may know more about room arrangement and its effect on learning than we do. Did you ever notice that most math rooms are set up in rows? Did you notice that teachers have the most success with kids in the first two rows and not much success with the kids in the back corners?

When the room is big enough, I arrange the desks in a circle. The ideal situation is tables. During instruction, the kids face into the circle and do their work on little whiteboard slates. You can monitor that they are practicing the skill that you are teaching. You pause and they show you the slates. You get immediate feedback. You can correct, expand, or change your instruction to fit what they are doing or not doing. Kind of like instant formative assessment and correction—if you are into those words. I like to call it "checking in."

Then, when it is time to do the "non-homework," the students turn to face the outside of the circle. Now they can only interact with two people—the student on their left and the one on their right. You can walk around outside the circumference of the circle. You can check on students, you can help them. You can watch the entire room all the time. Kids can only interact with you or the students next to them. Opportunities for students to mess about are sharply reduced. Your ability to see it and deal with it is greatly increased.

BUT most rooms are not big enough to arrange your class in a circle. So how can you get most of the circle's benefits? Your goal is to break up the corners and provide a way to get around the room easily while kids are doing their non-homework. Groups of four sitting at tables work almost as well as the circle. If you have to, you can use four old-fashioned desks stuck together. **Why four desks?**

Because two is not enough, three is just right, four will work, but five is just a party. Sadly, most rooms aren't big enough for groups of three so four seems to be the best compromise. Ideally, every student will have three tutors/helpers/colleagues working within the group.

PARTY! Groups of five or more tend to lose focus quickly.

At this point, some gurus will recommend that you carefully select the groups looking for a range of skills at each table and possibly some ethnic or cultural diversity. Don't get sucked in. If you put best friends too far apart, they will just shout across the room. If you stick enemies together, they will pick on each other, or worse, swearing and temper tantrums happen. You will be creating a lot of discipline stuff you don't want to deal with even up to physical fights.

I've tried a lot of different ways of assigning seats, letting them have a bit of say about the groups seems to work best. I let them fill out a "request" that lists one partner they would like in their group. Usually, Latisha and Karen will both have the same names on their papers, so it is easy to put them together. Then I make my best guess as to putting the pairs into foursomes. With minor adjustments that usually works pretty well—not perfect—but better than good enough.

Table groups are not as effective as the circle, but the tables break the class of thirty-two kids down into eight groups of four. With Pre-algebra kids, you really should have twenty-four students per class, but let's go with the worst-case scenario.

You can manage and monitor eight groups a lot easier than you can manage thirty-two separate individuals. In a short time, each group will develop its own little hierarchy. One student will be the leader, another will be the clown, one will be a worker bee, and the fourth person will be the seeker, always asking "why?"

Teacher on floor—not a good look! And who really needs a broken hip?

To truly exploit the power of the group, wander around while they do their non-homework. You will need some management of the backpacks. If the packs sit on the floor in the middle of everything, you will be tripping over them constantly. While you're walking around, notice who needs help with what. Kneel down and lend a hand. (If you are over seventy, figure out how you will get up before you kneel.)

Sometimes I will go from group to group and say, "I'll give you one free problem, which one do you want it to be?" If one student from another table has an excellent understanding, you can ask that person to move over and demonstrate. Since they know each other, they will probably do that. The whole class should be engaged and working for the entire time they are doing their non-homework. That's the point. You change the class dynamic from sitting, and maybe listening, to doing the work you demonstrated at the beginning of the class period.

Chapter 7

Struggling math students will give us as little as we will accept. D grades are quality killers—C grades are only a little better. We must like low-quality work since we have a special grade for it.

A grade of D is just an unspoken contract we have made with ourselves, our students, their parents, the district office, and the whole community. If it were written, the contract would say, **"I'm willing to pretend that you learned something if you're willing to pretend I taught you something."**

Would any of us settle for a 60 percent brake job? How about a painter who only completes 65 percent of the house and wants full payment? Of course not, none of us would be okay with that.

Yet kids, and their parents too, believe that they have a God-given and legal right to do lousy work at school and earn a D grade. We must believe it too since we are willing to grant high school credit for 60 percent of the work that was not very much, to begin with. We're so okay with D work that we reward it the same as A work. Both grades earn full credit toward graduation.

Nobody ever talks about the self-serving, almost sinister, reason that we're okay with D grades. We get to move the little chowder heads on to the next class, they're out of our hair. Your administrators are okay with low quality work because it keeps the parents happy, the phone stays quiet, and the district office folks don't come around meddling with instruction until the state results are published. Even then they don't really want to talk instruction, they just want to buy new books.

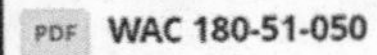

PDF **WAC 180-51-050**

High school credit—Definition.

As used in this chapter the term "high school credit" shall mean:

(1) Grades nine through twelve or the equivalent of a four-year high school program, or as otherwise provided in RCW 28A.230.090(4):

(a) Successful completion, as defined by written district policy, of courses taught to the state's learning standards. If there are no state-adopted learning standards for a subject, the local governing board, or its designee, shall determine learning standards for the successful completion of that subject; or

(b) Satisfactory demonstration by a student of proficiency/competency/mastery, as defined by written district policy, of the state's learning standards.

Blah, blah, blah. In Washington State, the law never addresses what level of work constitutes a particular grade. That decision is left to the district board of directors. Somewhere in that weighty tome called the Common School Manual they suggest a credit should be about 180 hours of work—but even that is squishy since the board can determine competencies in lieu of seat time. Kids do not have a legal right to earn a "D" unless the board says they do. As for the God-given right, I suggest we leave that the discussion to ordained math teachers.

Almost nobody with D grades ever passed the SBA or end-of-course exam. If you examine the scores, you will find that kids with a "C" average in math don't pass very often. So how do we fix this? It ain't easy, my friend!

Imagine my surprise. Luckily, a student who liked me, had a dad in tire business.

At Elton, where we developed the success strategies, we planned to give each student only as much credit as a student earned. We designed a continuous progress model. Students needed to learn everything we taught, and they needed to learn it well. Instead of A, B, C. D, F grades, they received a P on the report cards and transcripts, and they earned variable amounts of credit. If a student mastered half of the goals, she received .5 credit. Our target was mastery. We expected 90 percent on tests and 100 percent on daily work. You could say we were paying "piece-work wages." I frequently said, "We will pay you for all the good work you do, and we won't give you anything for low quality and incomplete junk."

It seemed like a great idea. It worked well at the alternative school, and it worked well for most of my Math Lab/AGA students. A few kids hated it, they hated the system and they hated me for creating it. I had trampled on their sacred traditions. I had violated the unwritten, sacred contract of the contract of the "D."

Wednesday night before Thanksgiving, cold, foggy, damp, sun nearly down—I walked up to my little Dodge wagon. It looked funny. It was sitting low. Yeah. All four tires. One hole per tire. Jose or Rick, probably both, had pushed a narrow knife into the sidewall of each nearly new tire. They both despised our new grading method. The guys hated redoing low-quality work. They refused to come in at lunch or after school to correct test problems. The pair of them went to the counselors and principal and demanded that I follow the law. "He has to let us pass with a D. *I always get a* D. *It's the law." Funny thing, I wasn't very mad, maybe a little surprised. Nine years at a hardcore alternative school and nobody had ever touched my car.*

Kids are comfortable with failure. Success is scary. *There's more to the Elton story.*

The principal, Robert, called me into his office. "Jim, you know how you are giving partial credit in proportion to the amount of work a student completes?" I nodded. He continued. "We probably need to change that, it's causing a lot of extra work for the registrar." I knew that was BS.

I gave her alphabetized lists of student names, student numbers, and the amount of credit earned. All she had to do was drop into the editing function and change all the grades at once. Ten minutes tops. She could do it the same way I did at the alternative school when I was principal. She just didn't want to do it.

But I kept my mouth shut. I stayed silent and just looked at him. I stayed silent longer, he began to fidget. He broke first and said, "I thought this is the place where you would say, 'We're not running the school for the convenience of the registrar.'"

I did not say, "Robert, we both know the supe told you to give me anything I want so you are stuck with this grading system. I'm sorry you're a new guy and the secretaries are giving you grief." I did say, "I was pretty sure that if I gave you enough time, Robert, you would get there."

A lot of what we do in education is for the benefit of the organization, not kids.

It gets worse.

We completed the first year of the pilot class. Kids who had not passed a math class lately gained a mean average of 2.5 years of skills based on two different standardized tests. The median gain was 2.2 years. (One was the MAP and I have forgotten the other.) Five kids passed the state test, and all the rest made huge gains. One more year of work, and I felt confident that we could get most of them over the bar. Things were looking good.

*The assistant principal called me into his office. He looked somber. "Jim, you have been accused of inappropriately touching a student, you have the **right**, blah, blah, blah, **union**, blah blah **investigation**, blah blah."*

"Don, this is bullshit! Before you tell me a single detail, I can tell you that this is all Lance, the guy who doesn't want to do B *work. He hates doing more than the crappy 60 percent minimum. Which girl did he put up to this?" I didn't get angry about the tires, but I was truly pissed about this allegation and called my lawyer from Don's phone. (I had a lawyer from my work as an alternative school principal. Some jackass was always threatening to sue me, and the conversation usually changed when I handed the jerk a card and said, "Have your lawyer call my lawyer.")*

The next day, the principal met with the girl and her parents. He felt that the mom suspected that something wasn't right about her daughter's story. Two weeks had passed before disclosure, and the daughter kept varying the account.

After an interview with the student and parents, the matter dropped. I don't know if anyone even mentioned my lawyer.

If there were a moral to the story, I guess it would be: It isn't easy to change the grading system. Kids are willing to destroy your life to preserve their right to do sloppy, incomplete work.

While that first introduction of Mastery Learning, ala Thomas Guskey, back in Elton School District was dreadfully dramatic, **it did**

work. Most students liked success. They learned to do quality work. Test scores shot up. The next year nobody blinked about the mastery grading model.

Fortunately, during the ten years of development, I learned how to ease into grading strategies that support student success.

Ten years later, I was working with high school seniors who had not passed the state math exam. We used the grading scale that made 95 percent an A. The B grade was 85 percent, and anything lower was I, incomplete. Every single one of our seventy-two students in three classes passed the class and passed the state test. When hired, I promised the administration that we could get 80 percent of the kids to pass. I secretly believed the pass rate would be 90 or 95 percent. I was a little shocked that we hit 100 percent. I must add that there was a little luck involved with the four students who made it by one point. If we stop rewarding low-quality work, we will get better results.

My principal, Nick, did receive one call that he described as "90 percent question and 10 percent complaint." Nick understood the mastery methods and told the parent, "Why don't we wait and see how Cindy does before we lean on Mr. Slosson?" Cindy turned in her missing assignments and got an *A* for the class. She also passed the SBA and the ACT. Since the sixth grade, she had never earned a math grade higher than a low C. But then, nobody ever insisted that she do quality work.

If we think about how grades normally work, we don't see much stress on quality for every student. We start with a chapter in our math book. We go through the ideas and concepts. Some students pay attention while many screw around. Kids may or may not do their homework—mostly the ones that don't need extra practice complete homework. Teachers may or may not do a few formative assessments (quizzes). They may or may not modify their instruction based on the formative assessments—usually not. Then we test students.

Many of our struggling students get lousy grades on tests. Some forward-thinking schools may try remediation for kids with low scores, the kids may or may not come in for help—mostly the kids never show up. Finally, soul-crushing apathy wears us down, we give up, and ready or not, we move on to the next unit.

With each new unit, the low-achieving students fall farther behind. They can usually eke out a D in the first semester. As the year continues into the second semester, kids realize they are hopelessly behind, they know failure is guaranteed. They quit trying. They know they will be repeating the whole class next year since they won't receive any credit for the tiny amount they did learn. They become a major pain in the class. They do no work. They don't even pretend to listen. They skip class.

The following year, we waste valuable resources starting over when we put the kids back in the same class. Then we teach them the same way we did before. We get the same results. The only difference in the second year of the same class is that they give up sooner. In a way, we are rewarded by the students' failure.

Over time, we get more math teachers in our school so we can do the same thing the following year to our ever-increasing pool of low achievers.

In manufacturing, we call this model the "scrap and rework bin." It is one of the biggest hallmarks of a very low-quality system. In business, low quality costs money. In education, low quality costs time, money, and student self-esteem.

We need to change the dynamic of low quality. It is not logical to expect that doing the same thing next year will give us better results than we got this year.

But you will need thoughtful change if you want a new standard of high quality for all students. My seniors were willing to accept changes to a high-quality system because they wanted to graduate, and I learned how to make finesse the process of changing grading systems.

What happened in those ten years?

- I learned how to get parents on my side early in the process. I explained that the student's only chance to pass was to learn to do quality work. Every parent received a copy of the class rules and a detailed explanation of the grading plan.

- We invited parents to an evening meeting during the second week of school, so students could show their parents how things worked.
- I supplied the counselors and administrators with a list of talking points that bolstered our plan.
- I shored up administrator and counselor courage with short, but frequent progress reports on the progress of the whole class and individual students who were falling behind.
- Provided weekly printed progress reports to kids and reminded parents to check the online grades.
- Scheduled make-up days twice a month, so kids could redo unsatisfactory work and deal with missing assignments.
- Changed grading and progress reports to a point system instead of weighted percentages, and carefully explained how the point system works.
- Leveraged the daily work scoring so that students had to do quality work to pass. The change was so subtle that nobody even noticed it. I realized that a high standard of quality work on daily assignments was nearly as powerful as my old Elton system. (See Grading Daily Assignments, Chapter 11.)

To ease into a higher-quality instructional model, start by changing the weighted scoring system that is built into your school's grading program. Turn it off. You know what I'm talking about. Test scores are 80 percent of your grade. Homework and class work counts for 15 percent of the grade, and participation is 5 percent of the grade. That gibberish gets all jumbled together on a progress report and somehow a letter grade magically pops out.

Forget all that. Kids don't really know what it means, parents don't understand it, and when teachers try to explain the system, it sounds like nonsense. Simple is better.

Just go with points. I usually make my class worth 1,000 points. Tests count for 800 points. All other assignments amount to 200 points. Those points comply with department/district policy, you have 80 percent for summative evaluations and 20 percent for assignments, and no points for teacher-pleasing behaviors. Every assignment matters. Every test matters. At the bottom of my progress reports, I include the comment, *"Please notice the number of points your student has earned to date. Students must earn 600 points to pass our class and receive credit. Students on track to pass this class should have earned 250 points as of today. Students receive full credit for make-up work even if turned in late."*

Keep it simple. Kids can earn 800 points on tests and 200 points on daily assignments. Everybody gets that. A parent has only to look at the progress report and know exactly what's missing and how it affects the grade. See a detailed example of the system in the next section devoted to grading assignments. There is a work-around if **they** won't let you change to points.

Each time you enter a new item on the grade program matrix, you must tell the computer if it is an assignment or a test or whatever. (There is a vast group of categories.) In my last sub assignment **they**, the district office folks, wouldn't let me use points. I don't think they trusted their teachers to stick with the system. They locked it down with weighted percentages. So, I just listed tests as assignments and gave them a 100-point value. Daily assignments had a five-point value. The ratio was still 80:20. I've been a prisoner longer than they've been guards.

You may balk at accepting late work. You might believe that penalties for late work incentivize kids to get stuff done on time, but you are wrong! Penalties are a punishing disincentive, and worst of all, penalties can create a situation where a student has no chance of passing the class even if they do A work the rest of the semester.

Even late work must be done completely (or nearly so) and correctly because it is important work, and it is work worth doing. Each of your assignments is worthwhile, and an important part of preparing for the test. It logically follows that every assignment must be completed late or not. Kids will buy this explanation if they get full credit for the work. It's a hard sell when late work earns reduced points.

What is the downside of all this simplicity? To make it work, teachers must be organized. Simplicity starts with good planning and good planning begins with "keeping the end in mind" (Covey) before the class starts in the semester.

Here is another instructional model that can reduce failure—something we tried at Elton. I was working part-time and didn't have many kids in my seventh period, the one after school. We were using the old APEX computer-assisted instruction, and I had an assistant that did all the computer work. I just worked with kids. Since we had space, we scooped up about twenty kids who were getting a D *or* F *grade in Algebra I. These kids were a self-selected group, not a random group because they were willing to stay after school. They weren't forced to be there. The kids stayed in their regular Algebra I class and did the APEX in addition.*

The students would spend thirty minutes on the computer individually working on the areas they didn't understand during the first semester. It turned out that most of them had a few specific things they missed. Those few elements had caused their failure, and the computer made it easy to find and to focus on the problem areas.

The other thirty minutes of class were worked on the second-semester homework they didn't comprehend. Remember these kids volunteered, they probably would have done the homework if they knew how to do it.

All twenty of them saved their first-semester credit, and nineteen of them passed second semester. They passed the end-of-course test, and they passed the final test.

So let's talk tests.

Chapter 8

Teach What You Test and Test What You Teach
If you aren't teaching the test, you're doing it wrong (or it's the wrong test).
And some rolling review is a good idea too.

Pour another beverage, and let's think about tests. Why should we consider tests before we've talked about assignments or instruction? Simple! Your instructional goals should drive the content of your tests, and your tests should dictate the content of your assignments and your daily instruction.

The test is our destination. Writing the final first is, as Stephen Covey said, "Beginning with the end in mind." Your unit tests and your final test should include an example of every standard you want kids to master, that's every skill you covered since the beginning of the year. If you incorporated rolling review into each assignment, the kids can demonstrate everything they learned and practiced.

You already know that the kids' calculating skills suck. There is no need to punish them further for something we (all of us in education) should have dealt with earlier. Let them use the calculators unless you somehow managed to teach them the math facts they should have mastered in fourth and fifth grades.

The test items should highlight the students' ability to use the math concepts you taught them.

1	2	3	4	5	6	7	8	9
2	4	6	8	10	12	14	16	18
3	6	9	12	15	18	21	24	27
4	8	12	16	20	24	28	32	36
5	10	15	20	25	30	35	40	45
6	12	18	24	30	?	?	?	?
7	14	21	28	35	?	?	?	?
8	16	24	32	40	?	?	?	?
9	18	27	36	45	?	?	?	?

I've tried many times with groups students to help them learn multiplication facts. Never had much luck. Kids are emotionally and intellectually resistant to working on elementary skills..

Some tips for writing test items:

- Include a few items that provide room to show and work and give the student the answer. That way, the test item becomes a "show me the steps" kind of problem.
- It's okay to give kids a set of hints on the test. For example, I like to include the point slope formula, the slope formula, standard formula, and slope intercept formula in a box on the linear systems test. The formulae just sit there with no labels, but they work great as prompts—some folks might call them hints.
- Let the students write their own equation and then draw the graph. "No, you may not use y = x."
- Let the students make up a word problem to fit the equation you provide.
- Try to use numbers 2, 3, 4, 5, 6, 8, 10, and even multiples of those numbers such as 20, 15, 35. Some kids can do those numbers without a calculator. For those who need calculators, they are more likely to get correct answers.

- Keep the common fractions simple such as 1/2, 1/3, 3/4. Avoid common fractions with larger prime numbers in the denominator or numerator. A number like 19/31 puts the student's attention on the wrong part of the problem.
- Leave plenty of room to show work.
- Leave blanks for the answers on the right side of the test, it speeds up grading.
- Include a few "rolling review" questions from previous units.

I like to write the unit test before I design my materials and plan the instruction. The first time I used success strategies, I only wrote the final and the pretest/study guide. The second time I designed a course, I spent most of the summer writing the final and all the unit tests. I also completed writing all of the daily assignments for the first three units before school started. Sixty and seventy-hour weeks were the normal that year. I didn't bother writing the practice and make-up tests until we worked on the units. You can do it in less time. If you're working on Pre-algebra, you can start with my stuff and improve it as you go along. The Math Lab/AGA materials include a practice test and real test after every ten lessons.

When I taught Geometry, I used the math department's assignments, but rewrote the work into daily graded lessons. I found that I could do about two an evening. The assignments were pretty simple, but remember every student completed that assignment with 100 percent accuracy in fifty minutes. The highly capable kids helped other students in their table groups and still had a couple of minutes to mess with their cell phones.

Pie are square. No, you fool, pie are round. Cobbler are square. That joke never gets old!

In your own work, use math department or publisher-made tests as a starting point, but there's a problem.

Those exams assume that the published pacing guide makes sense for your kids. It doesn't make sense. That's why they're in your class instead of a regular class.

Example: *I subbed for a Pre-algebra class, long term, starting in mid-January. Messed up situation. They suffered through five different subs in three months. One guy quit after two days. Some days there was no sub at all—other teachers grudgingly covered the class. Before leaving in mid-March, I made it my mission to ensure that everybody in both classes could at least work with circles; that's a fourth or fifth-grade skill. We worked with "C and D connect by 3 and radius looks like half to me" for six consecutive days until all of them had it. It took four of the six days for them to see and understand that the diameter is twice as long as a radius. Each of those days we needed to review the notion that three diameters equal one circumference. No pacing guide on earth assumes that it would take fourteen-year-old kids more than a week to relearn circles. We would still be there if we insisted that Pi equals 3.14, plus it doesn't rhyme very well.*

If you want to have some real fun with "real" math teachers, discuss the value of Pi; they go crazy. Using 3 instead of 3.14 is only a 4.5 percent error and plenty close enough when a struggling student is trying to memorize the ratio. Students can always switch to 3.14 when they have the concept in their heads and a calculator in their hands.

Once you have written the unit test, you'll need to make three or four more versions. But that can wait until you teach the unit. Since this is math, not history, you will only need to change the numbers inside the problems.

- Pre-test administered at the very beginning of the unit. This can also be a study guide the kids keep in their binders. John Hattie points out that a pre-test creates focus for students and increases learning. **But the truth is** that while I use pretests for most classes, I skip writing and administering pre-tests for Pre-algebra and older kids taking Algebra for the first time. Kids don't usually solve a single problem. They start the unit totally bummed, "See, I told you I'm dumb."
- Practice test. The one you give them a few days before the real final.

- The real test. The culminating, summative assessment—the final, final (sort of).
- The make-up test. This can also be the "Jackpot Final" if you are so inclined.

I suggest you hand-write the tests. Typing mathematics material is brutally slow. With some practice and some lined or graph paper that you make up on your computer, you can write out your tests quickly and neatly

But if you decide to do the pre-test, read on. Otherwise, skip down to **"I suppose."** This is how it can be done with more willing students who might know something about math.

When you write the test, leave enough room so that you can divide the answer space into two columns. In the first column, the students will write the pre-test answer. Administer the test. Most of them will do a miserable job. That's okay. The test exists only so you can find out what they already know (and don't know), and the students get a clear picture of what they will be learning. The students need to do a neat job because they will be referring to this test almost every day for the rest of the semester. (You might want to print this test on heavy paper, it is going to get a lot of use during the semester.)

When the pretests have been scored and returned to students, the pretests become study guides. The study guide makes up the first few pages of the students' in-class binder. When you teach a new idea or expand an old idea, ask students to open their binders and look at the problem they will be studying today. When you have completed the activity, students should write the correct answer—with all the steps—in the space to the right of the pre-test. In this way, they will have a comprehensive study guide for the unit tests and the final test.

I suppose—depending on your district—you could write the target learning number or core curriculum identifier or some such thing up in the corner of each question. Personally, I think it is unnecessary since I have never had a student or parent ask me if we were covering learning target 7-G.23. But such pseudo-precision makes for happy curriculum/teaching/learning directors, they feel they're in control.

They aren't. You are. It is a pain in the butt to look up all those little numbers, but you only need to do it once, and nobody is actually going to cross index the whole thing to see if you did it correctly anyway. It's Pre-algebra, nobody really cares!

Why do we need a practice test and a real test?

We need two tests because teenagers and pre-teens, all kids really, are quirky people. If you tell them there will be two chances to pass the test, almost all of them will blow off the first test. They won't study or review, and they won't take the first test seriously. BUT—what if we change the stakes?

What if you could get a high score on the practice test and avoid taking ***the real test***? What if you told them, "If you get 85 percent, pick a number, maybe 80 percent on the practice test, you don't have to do the real test on Friday. You can put on your headphones and listen to music, text your little heart out, or play games on your phone while other students take the real test.

If you do that, you will find that your students suddenly gain an interest in passing that "practice test." They will pay a lot more attention during in-class review. Some of the students will even try to sneak their binders out of the room so they can study on their own—God forbid! But seriously, don't let them take their binders home, they will lose them or forget to bring them back—remember they are mostly *P* learning style. They can photograph the pages on their phones if they have a mind to.

The make-up test has two functions.

Some students will legitimately miss the test. With a make-up test, we can be pretty sure that students did not have a friend sharing the "real test" with them beforehand. AND—now and then—we all get a class that thinks cheating (they call it helping) is okay. When you get a class like that, it's quite fun to pass out two versions of the same test—for added effect, you can print the two tests on different colors of paper.

And what, you should be asking, is the JACKPOT FINAL?

Sometimes you come across a youngster who hasn't really passed much of your class during the school year. Sometimes they were just messing around, sometimes they were incarcerated. Now and then, they have suffered through homelessness or a messy divorce or maybe even an assault. Yet for some reason, a particular kid has learned the math but not accumulated enough points to prove it.

What the heck, give the kid the end-of-course exam or comprehensive final. Let the youngster show you what he's got. If they pass with an 80 percent score, call it good enough, give them a P grade and move them on. There is just no sense in making this kid do the class over.

What if every kid just wants to do the jackpot final?

If they can pass the comprehensive, jackpot final with a score of 80% I'll give them credit for the whole class. It may feel like an A, but it's only a "P".

I guess you might ask yourself, "Why wouldn't that be okay?" If any class lends itself to the "challenge" model, math certainly fits the bill. Unlike social studies or writing, the proof of competence is objective—there is no subjectivity to math. Group interaction and discussion aren't vital components of the learning. Were we honest, we could just let kids test out. We just don't want to.

Be careful, don't question the status quo too much. Or you might start asking yourself why we require kids who fail just the second semester to redo the whole course. That could lead to questions about why we don't give math credit toward graduation for Pre-algebra. That could lead to questions about why everybody needs calculus when we know that less than 1 percent of our students will actually use calculus on the job, but somewhere north of 50 percent of them will use spreadsheets and databases on their jobs. But I digress. Let's focus on helping our little chowder heads get ready to pass Algebra I. If things are going well, probably best to keep your head down and stay off the district office radar.

Although the tests are supposed to be summative, we can recover a lot of learning with structured testing. Here is my schedule.

Unit testing takes about a week.

On **Monday,** we do a little review assignment together—one of the few times I spend most of a period in front of the class. Kids who are stuck can get help from their neighbors. Typically, I will give the whole class a minute or two to complete the problem. When they all look up, we put the solution on the board. This is a good opportunity to call on a few of your "smarty-pants" *kids to help out. When the review is complete, allow the kids to take the papers home. Mostly they won't look at them, but you did your best. *See smarty-pants points in the assignment section chapter.

Tuesday, the students do the practice test. I walk around the room with the answer key in my hand and grade tests as they finish. Occasionally, I will say something like, "Do you really want to go with that answer?" But that's all the hint they get from me; they want me to help, but I don't. About three fourths of the kids will score 80 percent or more. Technically, they don't have to take the "real" test on Thursday. They have already nailed down a B on the test.

On **Wednesday,** we autopsy the test. Most of the kids who got less than an A chose to participate because they can raise their grade to an A with no penalty if they take the "real" test. Students who already earned A grades could just mess around on their phones or read for pleasure, but they mostly work on assignments for other classes. That's why they earned an A. The first time I used success strategies, I only wrote the final and the pre-test/study guide. The second time I designed a course, I spent most of the summer writing the final and all the unit tests.

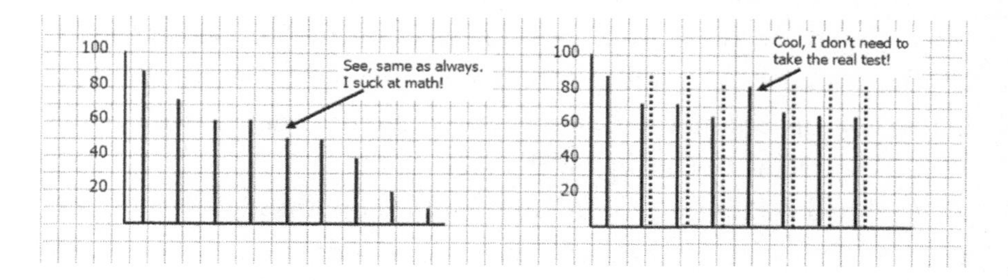

The two graphs illustrate the need to change testing practices. In a traditional model students do worse on every test until they finally fail. In the success model, interventions reduce failure.

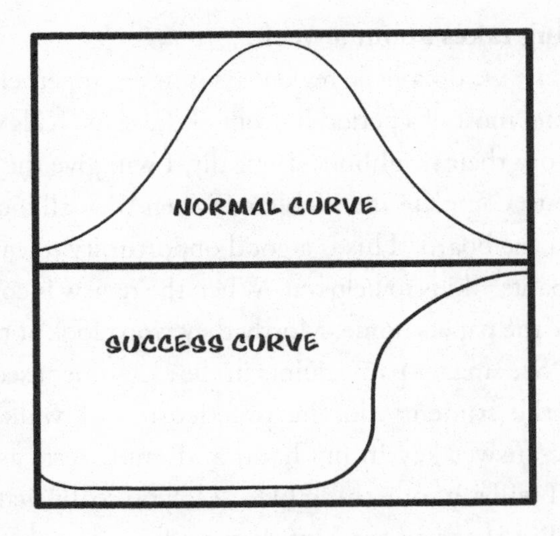

The upper curve is the standard that represents the random distribution of traits in a population. The "J" curve represents the result of intervention. I like my test curve to be nearly vertical. The bump on the left. Some kids are going to fail no matter what. My former student, Brad, who embraced failure could see I was in anguish over his "F" grade and consoled me. "It's okay Mr. Slosson. It's not your fault. I've gotten F grades from way better teachers than you."

Now it's **Thursday,** time for the real test. I walk around with the answer key and a highlighter. Tests are graded on the spot. I say little, but I have been known to point at a test item and sigh.

By **Friday,** almost everyone has passed. People who have grades less than 75 percent, there are always a few, are invited to join me before or after school by appointment for one-on-one work. I will reteach the items they missed and then ask them to correctly complete two or three parallel problems I make up on the spot.

Only a few kids take me up on this offer, and the highest test score they can receive is a B. Unfortunately, a few kids embrace failure like an old friend and don't bother to come in. For the students who do come for the one-on-one test repair, there is a wonderful unintended consequence.

I keep a few treats in my classroom, especially for students who come in for help. Usually just some cookies or chocolate bars. I only allow two students at a time, and I require that it be two students. The hallway door is always open, just common-sense precautions. We sit next to each other in desks. I stay in the middle, and the students work the problems on small whiteboard slates.

While they are relearning and demonstrating math problems, we engage in a bit of safe-topic, personal conversation, "How do you like the car you bought? Is your sister still mad at you? When will your mom return from deployment?"

What is the result of this non-math chit-chat? Kids start to believe that you are like them personally, "You know he's not as mean or strict as everybody says." When you invest in them, they will invest in you even if they don't give a hoot about math.

Some—most really—of my colleagues don't like this system. A few despise it. They think it's too much work, but really, it's more about a system they didn't experience as high school math students. They like tradition, and they don't like personal interaction.

While we're testing, most of the tests get corrected before the end of the period, and most days, all the tests are corrected and entered before I go home forty or so minutes after the school day ends. Once you get a handle on success strategies, life gets easier, it's less work, not more work.

A very few teachers resist the success testing regimen for a more sinister reason, punishment. It will never be articulated, but down in their hard, little, walnut-sized hearts, they believe the failing test grade serves the cause of social justice. For these folks, the failing grade is what the kids deserve, they are rude, impolite little jerks who don't appreciate me. What they say is, "an F is the logical consequence of laziness, and will teach students to pay attention in class and do their homework."

It won't teach failing students that lesson at all. It will teach them to hate math even more and to be a bigger pain in the butt when they take the class over again next year.

Now that we know, in some detail, what we what kids to learn, and how we plan to test knowledge and skill, let's talk about writing assignments.

Chapter 9

Every Day Is a Graded Day and Every Student Works Every Day

Creating daily assignments—the magic isn't in the materials but in your methods.

I called Brad down to my office. "Brad, your mom called, your brother can't pick you up today, you will need to ride the bus."

Brad started fidgeting, looking nervous, and sweating. The garbled words tumbled out, "I can't ride the bus. I've never ridden the bus. I won't know where to sit. I won't know where to get off the bus. What if I get off at the wrong stop? What if somebody talks to me? What if nobody talks to me?"

Brad stopped talking. I waited, then waited a bit more. Taking slow deep breaths that he started to mimic, I finally, said, "Brad, you robbed a liquor store at gunpoint when you were thirteen. I can't believe you are so stressed about riding a bus."

Brad didn't blink or pause, he stared straight at me, "Somebody showed me how to rob a store." There is a lot of wisdom in that little sentence.

We need to show our Pre-Algebra students how to work in class, it's a new experience for most of them. Highly employable young adults need to work through the whole period each day practicing what we show them. Call it non-homework if you like.

It appears simple, a single sheet of paper printed on both sides with about ten, maybe twelve, problems on each side. It is laid out so students can show their work—in fact they are required to show their work—all the time—every time—every day—on every problem.

Your struggling students are far more willing to complete a sheet of paper than they are likely to copy problems out of a book and transfer them to paper. One sheet of paper suits their *P* learning style. They aren't, in their words, "Overwhelmed by all that work and all those pages that just go on and on."

A kid with a *P* style is great at working on a handout because there is a narrow focus and a short window of time to complete the work. They don't need to take their book home. They don't need to remember to do the assignment. They don't have the opportunity to lose the assignment. You don't need to check books out. You don't have to check books in. Their parents don't have to pay for lost books. Hand-outs are a better deal for everybody except the textbook publishers. What about the cost of all that copying?

I did a spreadsheet in Elton and learned that the cost of all that copying was cheaper than buying and replacing books. And I didn't even count the improvement in goodwill when parents weren't stuck for the price of a lost book.

You are going to need to create each non-homework assignment from scratch. (Or maybe the ones I made in Elton will work for you.)

The first five problems should directly reflect the new skill you taught the students during your short burst of direct instruction. Problems six through ten will cover things that students have learned in the last week or so during the current unit. Problems eleven to fifteen direct the students to what we call **Rolling Review.** The last five problems expand on today's new instruction.

Students will complete the single sheet of paper assignment in class while you watch them and help them. Because you set the stage for student cooperation, and because you set up your room in eight small groups, it's easy for students to help each other. You have a classroom full of students who are also assistant teachers and tutors. The wonderful un-intended consequence of non-homework is that it keeps kids very busy and very focused leaving little time for messing about.

Unfortunately, these assignments didn't fall out of the sky or come from some textbook publisher. You are going to have to write them yourself. In many cases, you can be guided by the materials the rest of the department is using. That will cover problems one to five and sixteen to twenty on your assignment. BUT you are going to want to modify the publisher's work. You will want to change calculations so that they make sense for kids who don't know that 7 times 8 is 56. You are teaching math concepts and they can be learned with "kid-friendly" numbers. Those beginning and ending problems should directly address one or more questions kids will see in the unit test and the final test.

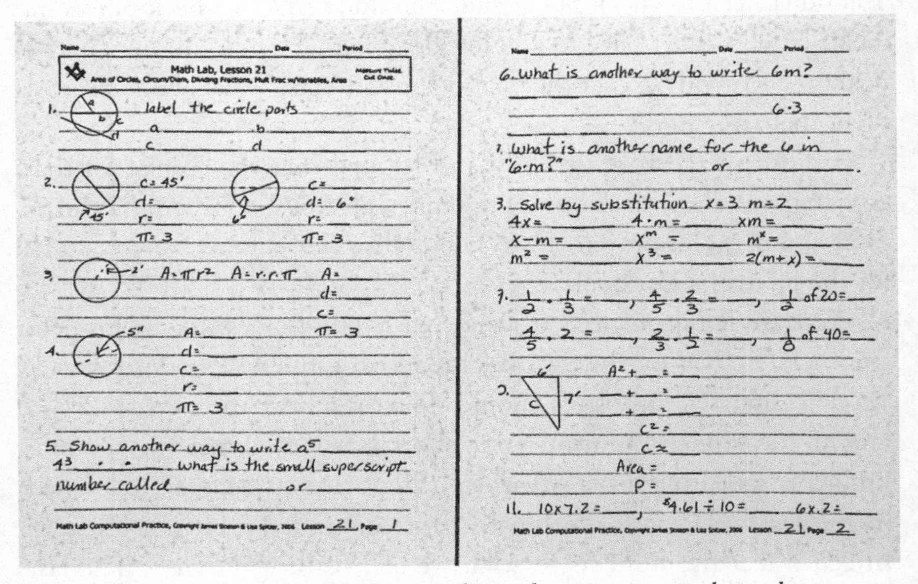
Name ______ Date ______ Period ______

Math Lab, Lesson 21
Area of Circles, Circum/Diam, Dividing Fractions, Mult Frac w/Variables, Area

1. label the circle parts
a ______ b ______
c ______ d ______

2. c= 45' d= ___ r= ___ π= 3 | c= ___ d= 6" r= ___ π= 3

3. 2' A= πr² A= r·r·π A= ___ d= ___ c= ___ π= 3

4. 5" A= ___ d= ___ c= ___ r= ___ π= 3

5. Show another way to write a⁵ ______
4³ ___ · ___ · ___ what is the small superscript number called ______ or ______

Math Lab Computational Practice, Lesson 21, Page 1

Name ______ Date ______ Period ______

6. What is another way to write 6m? ______
6·3 ______

7. What is another name for the 6 in "6·m?" ______ or ______.

8. Solve by substitution x=3 m=2
4x= ___ 4·m= ___ xm= ___
x−m= ___ xᵐ= ___ mˣ= ___
m²= ___ x³= ___ 2(m+x)= ___

9. 1/2 · 1/3 = ___, 4/5 · 2/3 = ___, 1/2 of 20= ___
4/5 · 2 = ___, 2/3 · 1/2 = ___, 1/8 of 40= ___

10. 6' 7' c A² + ___ = ___
___ + ___ = ___
___ + ___ = ___
c² = ___
c ≈ ___
Area = ___
P = ___

11. 10×7.2 = ___, $4.61 ÷ 10 = ___ 6×.2 = ___

Math Lab Computational Practice, Lesson 21, Page 2

This assignment would come early in the semester, perhaps during the third or fourth week of instruction. The new instruction focuses on circles. These assignments were written for 90-minute periods so there are four pages for each day. With 50-minute periods I do two pages (one sheet of paper) per class period. If you can get the kids to work hard, they will usually finish the sheet by the end of the period.

For problems eleven to fifteen, look back all the way to September. Select five problems from previous units that students completed weeks or even months ago. You can just flat out copy those. It's not like the kids will remember the answer.

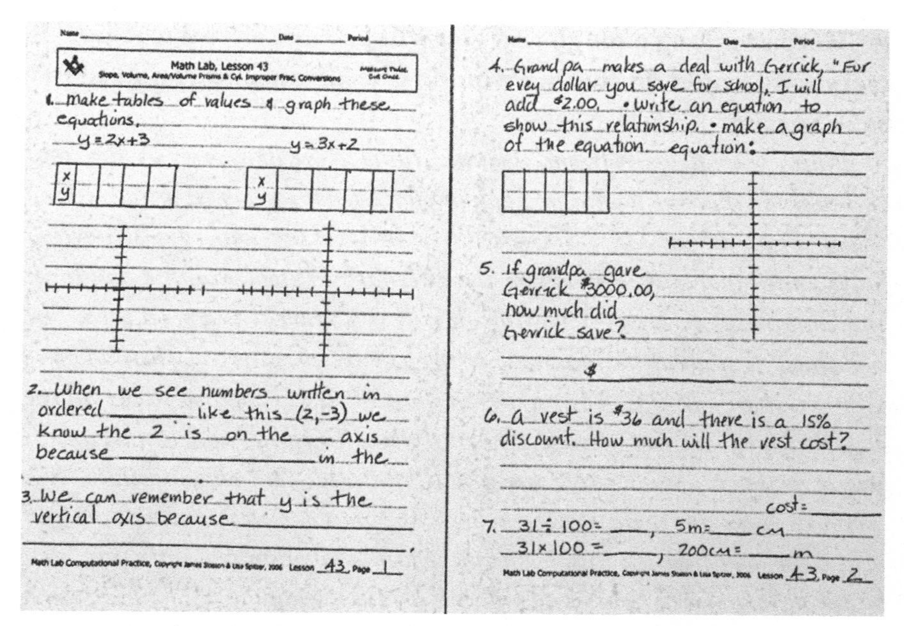
Name ________ Date ______ Period ______

Math Lab, Lesson 43
Slope, Volume, Area/Volume Prisms & Cyl. Improper Frac. Conversions

1. make tables of values & graph these equations.
y = 2x + 3   y = 3x + 2

x					
y					

x					
y					

2. When we see numbers written in ordered ______ like this (2, −3) we know the 2 is on the ___ axis because ____________ in the ______

3. We can remember that y is the vertical axis because ____________

Math Lab Computational Practice, Lesson 43, Page 1

Name ________ Date ______ Period ______

4. Grandpa makes a deal with Gerrick, "For evey dollar you save for school, I will add $2.00. • Write an equation to show this relationship. • make a graph of the equation. equation: ______

5. If grandpa gave Gerrick $3000.00, how much did Gerrick save?
$ ______

6. A vest is $36 and there is a 15% discount. How much will the vest cost?
cost: ______

7. 31 ÷ 100 = ____, 5m = ____ cm
31 × 100 = ____, 200cm = ____ m

Math Lab Computational Practice, Lesson 43, Page 2

This assignment will turn up somewhere around the 3rd or 4th quarter depending on the class. By this time we are working on 7th or 8th grade math skills.

This rolling review is key to higher scores on the class final and ultimately to passing scores on the state-mandated (or you could use the word required) assessment (or you could use the word test).

Yes! You are exactly right! This is a hell of a lot of work! That's why most people don't do it. But cheer up, you will only need to write about 150, maybe 160 assignments. After the first ten or fifteen assignments, you'll find that you can do it quickly. Don't bother trying to type them—it takes too long. Just hand write them on a standard form you create.

When I started, it took about an hour per assignment. With practice, you can get it down to about forty minutes. During conference weeks, I could knock out five or six work sheets a day. When you revise all the assignments (as you will because you kept notes about what worked and did not work), you will find that you can fix an assignment in about fifteen minutes.

Keep at it long enough and you will become a "real teacher." If everybody would do this, we wouldn't need you, and you wouldn't be so special.

Many years ago, my friend, Connie, came to me in my role as principal and told me that she would be applying for a teaching job in a snooty prep school. She had an "in." She was tired of working with our struggling thugs and impoverished moms in the alternative school. Connie had built an elaborate structure of reasoning on why this move was a good idea. She explained that wealthy kids had struggles and pressures of their own—different than our kids—but struggles none the less. I listened to her all the way through; said not a word. When she was finished, and I must admit this is really crappy, I patted her hand, and said, "It's okay, Connie. Not everyone is cut out to be a real teacher." She glared a mean stare at me and held it for over four seconds, seemed like forty. Then smiled and said, "You bastard! I wondered how you would try to talk me out of it." She didn't talk to me for a couple of days, but a few months later, when we were working on the master schedule, she tossed off an aside, "I'll take the support class—some folks seem to think I have 'the gift.'" Follow-up: Connie has spent the last twenty-plus years bringing the joys of literature to struggling middle school students. They love her enough that they will read just to please her. She has even convinced convicted felons that they should stand on a chair and swear the sacred oath of book borrowing.

Now that you have your assignments written, you'll need copies. If you are blessed with a copy center, you will want to cultivate a relationship with the people who do the copy work for you. I always drop by on Fridays with a doughnut or two—maybe a pastry. A birthday card or small gift on holidays is also a good idea. The copy center person makes your efforts possible.

If you must do the copying yourself, train a couple of student assistants. Take in about ten assignments at a time for printing. That will last you about two to two and half weeks or maybe three weeks. Give the copy center folks plenty of lead time. Unless it is the week before finals, a week should be enough. Avoid Monday mornings and Friday afternoons. That's when all the P-style teachers swarm the copy center. You want special treatment so a little consideration and organization will go a long way with overworked and underpaid classified workers. Diligently complete the work orders.

I ask the printers to copy in backward order. Print assignment ten first. Place it in the bottom of a paper carton. Then print assignment nine and place it next to ten on the bottom. Put a colored separator sheet on top and then assignments eight and seven, colored sheet, and so on. Now you have two weeks' worth of assignments neatly arranged in order. A week later, take the tests you will be using for that unit down to the copy center. You want to be the copy center's steady, pleasant customer—not their pain-in-the-butt customer. If student assistants do your copying, make sure they feel loved—a special treat now and then along the occasional Starbucks card will buy a lot of good will.

All the work you put into organizing these assignments pays off in other ways that save you time and effort. If you have two classes of twenty-five, print at least twenty percent extra assignments. That sounds excessive, but that will barely cover what you need. These kids are all P-style. They can lose papers without even leaving their seats. Store the extra assignments in a simple filing system. I always liked milk crates with hanging folders. When I couldn't do that, I used big bulldog fasteners and stored the extra assignments in a box. With everything organized, you can put responsibility for gathering lost or missing work back on the students.

If your kids miss a day of school, the students can go to the folders and find the assignments they need. When some parent lets you know that they want work ahead for a family trip—no fuss. The kids pick up the assignments they will need down in Disney Land or out hunting, or whatever. No need to hassle the kids about this. They probably won't do the assignments anyway, but now it's on the student and the parent—not on you. No parents complaining about your lack of cooperation, and that's always a good thing.

I don't let students take their assignments home. They just lose them. unfinished assignments stay in the room, on the shelf, inside each student's individual binder. Organize the binders by class and table group for quick, silent retrieval. Every ten days or so, we have a "catch-up day." Kids work on stuff they didn't turn in, do make-up tests, go to the library, or just sit and waste oxygen.

Bonus points

You can keep your more capable guys busy longer if you add a "smarty-pants" bonus question to your assignments. Just make up some longer, harder questions or maybe something divergent that requires some internet research. If a kid gets it done, give her a 6 on the assignment instead of a 5. It keeps them busy, it costs you nothing, and they feel good about the 110 percent on their progress report. They are going to get an A anyway, so everybody wins.

Kid loses an assignment?

No need to point out that they were not supposed to take it out of the room—just point to the files in the back corner. The students will want to make their lost work your problem. It needs to be the students' problem. They can look up their missing assignments by accessing the grading program on their phones, and I give them a hard copy report every two weeks. But if you must, you can print out a progress report when you have time. Students need to know which assignments are missing and then pull them out of the bins you provided, their problem, not your problem.

Chapter 10

An hour of teaching is an hour of learning.
It's not an hour of talking.
Fifty-five minutes goes fast.
Some folks think teaching math is a performance art.

The teacher stands at the front of the class using some modern-day substitute for the old-fashioned overhead projector. Teacher shows kids how to do homework. Homework gets collected or posted on Canvas—probably not graded. Teacher stays at overhead explaining new material. Teacher works a few more problems. Kids clap and cheer, and thunderous applause rocks the room. Sometimes, the kids stand. They gush accolades, "Oh Mr. Slosson, you are so smart. Thank you, thank you so much for being our teacher. The X method of factoring is so totally awesome." Yeah! Not! Back to reality. One third of kids start homework. Two thirds screw around. Teacher checks NCAA brackets. He is not going to win the pool. Damn!

Mr. Slosson receives a standing ovation.

Learning is not about you talking, demonstrating, and performing. It's about the students working hard to master the skills you briefly demonstrated.

But before we can analyze a learning period, we need to address the issue of "smarty-pants" kids.

Seems like every class contains a little cluster of "smarty pants" students. They know every answer to every question. They raise their hands right away, they always finish first and would happily dominate the class period if they could. You also have a group that hasn't raised a hand since third grade. That's why I don't let students volunteer to answer, I call on them so I can spread the joy around.

As you look at the sea of not-so-eager faces, try to ensure that the student you called on can answer the question. Save the easy questions for the kids who need help, but if they stumble, ask another student to assist. Make sure the helper talks to the student you called on and does not tell you, or the class, the answer. If you increase your wait time, reluctant students will give the class better answers. I do try to let each "smarty pants" kid answer one—but only one—question. They would like answer instantly, but I hold my hand up and make them wait.

Most teachers expect students to answer quickly. I think the average wait time is just a tiny bit under one second. There's a lot of good research that shows the quality and quantity of answers will improve if just wait three seconds. I like to put the question to the class and wait a bit before I call on a student by name. Most kids will test their own memories just in case I call on them. Some days, I must remind myself about wait time before we wade into the high success format and demonstrate the lesson.

A high success period should look like this.

Zero minutes into the period—fifty-five minutes to go. I like to greet students at the door and remind them what materials they will need for the day. I remind them the plan for the day is on the board. It is important to greet each student by name and make eye contact while you welcome them to class. This serves three important functions. It builds relationships with students, and it reaffirms to every student that they are welcome in your classroom. It reminds them that you oversee this space. The room is your academic home that you happily share with your students. And your greeting sets the tone that we are here to do work.

I am writing this as we enter the third year of the Covid pandemic, I hope that handshaking will come back into style because the physical contact enhances the greeting and gently establishes your position as their leader. During normal times, the handshaking ritual is the basis of an individual lesson in my class that includes practice and checks for understanding as students learn to greet each other as highly employable young adults. Some kids don't want physical contact, covid or no covid. Muslim females have cultural prohibitions, some kids just don't like touching. Honor that and settle for a few friendly words, a nod, and a smile.

While students enter the room, I tell them if they will need their binders. My students must keep all their unfinished work in the room and the binder is an important instructional tool. Each period has a separate color binder or a separate-colored name plate inserted in the spine. When students pick up the binders as they enter, we avoid the traffic jam in front of the shelves. Sometimes, a designated table group member picks up all four binders during the period.

Kids hopping up and down to sharpen pencils during your few minutes of instructional time breaks up the rhythm of the class. Nobody is allowed to sharpen a pencil during my class time, and they don't need to. A supply of sharpened pencils on a special holder sits right by the door. Need a new pencil? Quietly walk over and get one—make no fuss.

Two minutes into the class, fifty-three to go!

Every student has a binder. Every student has a sharp pencil, and we're ready to start today's lesson. Every student has a copy of today's assignment.

Hold on—let's take roll. I know the experts say that you are supposed to have an entry or "sponge activity" that will engage students while you take roll, but the experts don't have your class of struggling students. Taking roll is an opportunity to build relationships among the students and gain a sense of order and calmness.

At the beginning of the semester, we do one of the "naming activities" I mentioned earlier. When all the names are learned, we focus on other relationship-building tasks.

Eight minutes into the class—forty-seven to go!

Most schools require that you enter the attendance into the computer yourself. It usually takes me about four minutes to enter the attendance. Before I sit down at the computer, I direct the students to look at a particular question posted on the board. "You guys look at question 17 while I enter the attendance." During that time, a student passes out the old assignments that need to be fixed. On make-up days, the students pass out the printed progress reports. This is also the time I let my students check their phones for those vital social-media messages that are the primary focus of teenage lives.

Twelve minutes into class—forty-three to go!

I go to my teaching station which may be at the board, or an overhead, or sometimes, I just pick up a small whiteboard and walk around. Sometimes, I stand at the back of the room and ask students to take my place. Normally, I demonstrate how to do today's work. I do the first problem on the assignment. The students follow along and do the problem on their work sheet. I walk around and check to see if they understand. Often, we will do the second or third problem too. I ask selected students to help others or clarify questions. When it looks like we're ready, I bark out, "Get to work."

While the kids work on the problems, I walk around and check to make sure the students can do today's tasks. I notice who is struggling and who has this skill down cold. I assign table partners to help each other. The kids have the first two problems done. **My direct instruction used up ten minutes or less.**

Everyone is working because this assignment is due in thirty-three minutes. I keep checking the first two or three problems. We want everyone to have two good samples to refer to while they work.

I need to interrupt our period timeline here to talk about teaching signals. When my table groups are working smoothly, it's hard to get them to stop and listen if I need to address the whole class instead of just one table. Lately, I have been using a little bell attached to the wall as my teaching signal. It's about three inches in diameter. Hit it once. It makes a nice clear ring. All the students stop and focus on me. I share my wisdom, and they go back to work. Over the years, I have used buzzers controlled by a remote complete with flashing light, a small gong, and a gavel. It beats raising your voice and adds a nice patina to your schtick.

Twenty-two minutes into the class—thirty-three to go

For the next twenty-five minutes—maybe more—I am walking around the class picking my way through backpacks and long teenage legs while monitoring progress. Students finish the remaining eighteen problems. I carry a blank copy of the assignment and the answer key. Sometimes, I will stop at a table group and say something like, "I will give you a free problem, which one do you want me to do with you?"

I carry a highlighter. If a student completes the assignment, I will mark it with a "5." I look up at the clock and remind them, "Eleven minutes to go." The rule is that when an entire table group is finished, the students in the group can look at their phones. They push hard. These assignments are due at the end of the period. A few kids ask the same question every day. I give them the same answer every day. "No, you can't take this home, get to work!" Sometimes the answer varies a bit, "No, you can't take it home, you know you'll just lose it." They usually agree.

"I need a pencil." "Got it." I need a triangle." "Right here!"… The Filson Cruiser's vest is no longer available, and it was pricey—but there are cheaper things that will work just as well.

And how do I carry all that stuff around? I wear a nice cruiser's vest. With plenty of pockets, it allows me to carry pencils, pens, highlighters, and small drawing instruments, a couple of protractors and triangles, and even a pad of hall passes.

Forty-three minutes into the class—twelve to go

More than half of the kids have finished, their papers are graded. They are maintaining vital social contacts. A few of them work on assignments for other classes. The other students are still pushing to finish. Questions six through fifteen are easily completed because they are rolling review of earlier work. With a little help from each other, almost every student finishes. I keep walking around helping and grading. I do not allow students to hand me finished assignments. I will just lose them as I move around the room helping kids.

Fifty-three minutes into the class—two to go

I announce, "Clean it up!" Once a shop teacher, always a shop teacher. They put the binders away. They drop their papers into the turn in basket or hand them to me. I stand by the door to make sure they stay out of the "death zone" by the door and remind them to return their pencils—with practice, most of them actually put pencils back in the holder.

Fifty-five minutes into the class—zero to go

The next period comes in. Repeat—the day goes fast.

I can finish grading the remaining assignments and enter them into the computer during prep period and lunch. After school, I pull out the assignments for tomorrow and plan that lesson. Planning doesn't take long, I planned this lesson when I wrote it two years ago and I revised it last year. I kept notes, it's coming back to me because I keep the lesson keys with the notes on the margins in a binder.

If I have a lot of time, I will fill out my next copy center request. I usually leave forty-five to sixty minutes after the school day ends. Today is in the books, and I am ready for tomorrow. I take nothing home except a mild case of exhaustion and my sheet metal lunch pail.

Minus forty minutes—end of my work day.

I drive home cherishing the silence and thankful for such a wonderful job. Each day of our careers, we teachers get to make the world a tiny bit better than it was yesterday.

Chapter 11

The Magic of Grading Daily Assignments. And more magic—you don't need to take papers home.

Suppose, I'm saying let's just pretend here for a moment, that you had a magic wand that could increase the quality and completeness of the work your kids do every day? As it turns out, you already possess such a fantastical device, you need only activate its latent powers.

It's called a highlighter, and like Excalibur or the Singing Sword, it's a powerful weapon against the evils of crappy, sloppy, incomplete work.

Every kid working every day on every assignment is the engine that drives the success strategies. When every student works the whole period, discipline issues go down and test scores go up. Daily grades matter. They make up 20 percent of a student's semester grade. And you totally control the quality of the students' daily work. Your highlighter is powerful!

For the magic to work, you need to tell them, "Complete every daily assignment, and you raise your final grade two levels." Abracadabra is optional.

Unleash the power of your Highliter!

You might want to make up a little table to show kids how things work.

Would it not be swell if you could grade all your papers and still leave school thirty to sixty minutes after the end of classes? You can.

As you walk around the room motivating, monitoring, cheering, and helping, you can grade completed assignments. Good work earns five points on a daily assignment. When every assignment is completed every day, and the work is done well, a student is on track to raise their grade two levels. A poorly done assignment only earns one point; forget scores of 2, 3, or 4. They're quality killers that dilute the push for good work—just a D in disguise. A score of 1 in your grading program indicates that the student didn't do much. So effectively, you have created a grading system that classifies daily work as an A (100 percent) or an F (20 percent). You are pushing the kids toward quality a tiny bit at a time, and because it's only five points per day, they don't even notice the shift.

I move constantly during the seatwork portion of the period. While checking work, I hold a pencil, a highlighter, and a whiteboard pen in my left hand. If the assignment is complete and correct, I mark a 5 on the paper, the student can use the rest of class time to do other things including mess with their phone. If a few problems need work, I don't mark anything on the paper. I just point out the mistake, and the student continues to work.

How much do daily points matter?

A bunch ± 2 whole grades

Test Points	Sem Grade w/o daily	Max daily work	Total Points	Semester Grade
800	80% - B	+ 200	1000	A+
720	72% - C	+ 200	920	A
640	64% - D	+200	840	B
560	56% - F	+200	760	C
480	48% -F	+200	680	D+

Daily work matters so much that we need to give kids opportunities to make up missing work ***without penalties.***

When kids won't work, I sometimes write on the paper a comment such as, "Refused to work." Or I might scrawl, "Would not get off phone." A youngster just sitting doing nothing, wasting our oxygen also merits a 1 marked at the top of the *paper. Sometimes the kid will get back to work—sometimes not. You can lead a student to water, but you can't make a student marry a fish.*

But comment or not, a score of 1 is only 20 percent of a score of 5. Out of necessity, most students will fix or redo the poor assignments. For most kids, this subtle transition to quality only takes a few weeks, and since they don't notice the change, they don't slash your tires or make false allegations about your behavior.

If a youngster is absent and doesn't do the work—even if legitimately excused—the kid has a recorded score of 0 in the grade book. Zeros catch a lot more attention from students and parents than a blank grade or an asterisk. When zeros get noticed by kids or parents, assignments tend to be completed. I don't excuse students form assignments because of legitimate illness. Sick or not, the student has skills to learn and practice to complete. You will get a few phone calls about this, but provide a cheerful explanation that students to learn everything and a reminder that kids can make up everything for full credit. Most parents will support the system. Most parents like the grade column labeled "device" or "phone." That's why you need a whiteboard pen in your pocket. Use that pen to put dots on the paper when a student is messing with their phone instead of working. It takes less time to make a dot than to write a comment. That dot earns a minus point in the device column. Each dot is -1 point in that column. When the youngster accumulates negative five points, they receive an email and maybe even a call to parents.

Students want me to collect their papers as soon as a 5 gets marked on the top. I won't take the papers. I tend to lose things all the time, but it gets worse when I'm moving around the room helping students. I bagged turn in-baskets and turn-in boxes a few years ago. Now the students hand me their papers as they leave the room. I clip the finished assignments together by period. Most kids finish before the period ends. Most days, I can finish grading the assignments and enter them into the computer during lunch and prep period and right after school. I seldom take any work home.

I don't return graded papers unless students need to redo them or fix a couple of problems. I just drop them in a filing cabinet drawer. The system prevents slackers from just copying work from another student. Now and then, rarely, I will make a mistake in scoring a paper. Sometimes, two pages stick together. Even though it's messy, this "archive" helps reduce the I-swear-I-turned-that-in drama.

If a student wants to contest my record keeping, we go through a couple of steps. First, the kid digs through the backpack, which is where they usually find it. If it's not there, we look in the "re-do" and "no-name folders." If we still haven't found the paper, I let them dig through the archive—storage drawer. Normally, the stored papers get dumped at the end of every grading period after I have grade proof sheets filed away.

A few months into the class, most parents and kids comment, "This works, why don't other teachers use this system?" Why indeed?

Chapter 12

Classroom Procedures and Management, Reducing Fuss
Routine things should be done routinely.

I dropped by a middle school and observed two separate eighth-grade classrooms. In class A, the teacher said, "Get your workbooks." Twenty-eight separate students scrambled around tables pushing and shoving to get to a disorganized, sloppy pile of workbooks stashed on multiple bookshelves. The whole grueling process took up more than five minutes of instructional time. My teeth hurt just from watching. Once the kids had their books, it took another few minutes to get the kids to settle back down. I was thinking to myself, "This is late November, this mess will continue every day until the end of the year."

Classroom B could not have been more different. When the teacher said, "Please have your group person number two get the books for your table," eight students quietly walked over to a single bookcase and picked up the correct four books for their group. The books were organized on one of five shelves—one shelf per class. There was a diagonal, colored stripe across the backs of the books so they could easily be identified (and replaced) in order. Each book was marked with the table group letter and the student number. The whole process took less than two minutes and created no commotion.

Good organization saves learning time and reduces chaos with the attendant discipline problems. By the end of the year, teacher B will have saved at least 600 minutes (ten hours) of instructional time, and countless migraine headaches. As you might guess, teacher B invested time in organizing the class and teaching (training?) kids to be efficient learners.

How might you use these same techniques to lighten your teaching day? If you can free up some time, visit a well-run fourth-grade classroom. I am not suggesting that you should have kids line up along the wall but observe how elementary teachers transition from one activity to another task. It's almost seamless. Check out the way students turn in and hand back papers and all the other little diddly tasks that suck up time and effort. Pay attention to room arrangement and signage. None of this happened by accident. Elementary teachers spend time and effort teaching behavior and organizing the classroom so they can spend more time teaching content.

Kids thrive on a predictable routine. Students may chafe against rules and procedures, but if you catch them in a moment of honesty, they appreciate a well-run, organized classroom. Try to start and end every period the same way every day. Your methodical approach to greeting students at the beginning and sending them on to their next class at the end of the period should bookend the hour they spend with you.

I find it useful to put the daily schedule on the board in two ways. Most high schools where I work run the "regular" schedule about fifty percent of the time. Lately, I have been in schools that use ten schedules in addition to the regular one. The most efficient classrooms hang a sheet protector or a clip board on the wall. Today's schedule is displayed out front with the others behind. Now really, this is mostly for your benefit. Kids who can't remember that 56 is the product of 7 x 8 can easily remember the release time for every class on every schedule.

The other schedule I post is the "whadawedointoday?" schedule that usually includes something like: "Roll, Interview (Madison), Asmt 14b, Prog Rpts, Catchup tomorrow."

Since there is no published correlation between Chrome Books, Canvas, Google classroom, and all the other non-magical crap we inflict on ourselves, and because I don't like devices anyway, my classes just use good old paper—still the most reliable information storage and retrieval system known to humans. Paper syncs up with everything and the system never crashes. But without digital storage, you create a lot of paper that can bury you unless you have a plan to deal with it.

Ah—good old paper. Always charged, always syncs.
It can bury you, or you can master it.

A couple of hints for handling assignments:

Keep the assignments in the room. I get one-inch binders for all the kids, if you can do it, a different color for each class is best. I cut out strips of index paper that fit the spine of the binder. Students write their names on the papers and slide them into the spines. Each class set of binders is kept on a different shelf in the bookcase.

When students enter the room, they follow our routine and pick up their binders. At the end of the period, any work that is not completed stays in the binder. Don't let them take the worksheets home, you will never see those assignments again.

If you can't get different colored binders, use different colors of paper on the tab that slides into the spine. In addition to unfinished assignments, the binders are an ideal place for the students to keep reference notes.

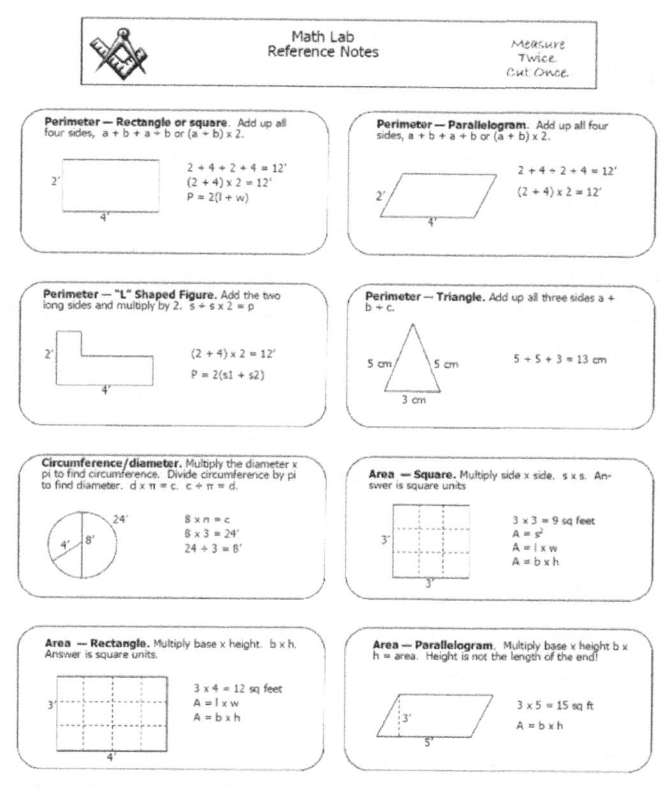

The Math Lab/AGA binders contain six pages of these notes. I usually print them on index stock so they will last the year in the binder.

While the notes are not textbooks, they are a handy way for kids to look up "stuff" they forget. When I am moving around the room helping students, I generally ask them to open their notes and I point out how they could have found the answer. Sometimes, they start to do that on their own—sometimes. Handing out worksheets can vary. In the last few years, kids get the daily assignment from me as they enter the room, and I hand it to them. When the class starts, I just leave a little pile on a table by the door. Late arrivals pick one up as they enter.

Turned-in papers can be handled in different ways. When kids were stealing assignments to copy, I had a locked box with a slot on the top. That worked great except that I spent a lot of time sorting and arranging stuff so I could score the assignments. Some teachers like tiers of wire baskets, one basket for each period.

Lately, I just stand by the door and collect the work as the kids leave the room. Most of the papers have already been scored. I put the unscored papers on the top of the pile and scored papers on the bottom. By the time the students are gone, I have a neat pile correctly oriented and clipped together. Grade entry is easy. A bulldog fastener holds them together in a neat pile, and I set them in my one basket next to my computer with a little practice grading goes quickly. You will need some little system for the few papers you are going to hold or return.

I use three folders, red for rework, blue for cold storage, and plain manila for papers without names. You could use baskets to store assignments. When I did my long-term sub stint, we came up with the idea for a folder labeled, "Cold Storage." Since the kids didn't have binders set up the way I prefer, we kept the unfinished assignments in the blue cold storage folder. Kids liked that folder. They, too, know that when an incomplete assignment goes into their backpack, they will never see it again.

As mentioned earlier, graded papers are not returned. I just drop them in a filing cabinet draw, sort of in chronological order. The top of the pile is newer work, the bottom of the pile is old work.

Pencils? Not an issue. I have a special pencil holder I made. Kids who need a sharpened pencil "borrow" one. I lose about five or six pencils a bad day. Most days just one or two go missing.

Kids are not allowed to sharpen pencils during class, they just borrow one of mine. My student assistant sharpens pencils and replenishes the holder once a day. I keep a few pencils on my person in the vest.

Calculators? I keep a few in a box, but mostly, the kids just use the calculator on their phones. For the math, we do it makes more sense, and it takes less keystrokes to find a square or cube root. I do carry a calculator or two in my vest for those few kids that don't have a device.

Bathroom? *"Don't ask me, and don't tell me what you're going to do when you get there. Just put your name on the sign-in, sign-out sheet, pick up the pass, and leave with no fuss. Return with no fuss too."* That's how highly employable young adults do it. Now and then, I have a youngster who abuses the system. They must ask me before leaving, but we have a secret, silent signal. I just nod if the request is granted.

If you set up a system for students to access previous worksheets, let your TA take charge of replenishing the folders.

I know students can access their grades and find missing assignments with their devices, but I like to go with a printed progress report every time we spend the period on a "Catch-up Day." Saves time, saves confusion. We don't worry about internet connections or charged batteries.

Look for opportunities to streamline your procedures. It is totally worth an hour or two of your time if you can prevent four minutes of aggravation every day. Four minutes, times 5 periods, times 180 days equals 3,600 minutes a year—divided by 60 minutes equals 60 hours a year which is a 3,000 percent return on your investment. But it's actually more than 3,000 percent because it compounds daily—sort of.

Chapter 13

Discipline is about teaching correct behavior—not just punishment.
If they won't sit down and work, you can't teach them very much.
Some stuff they didn't tell you in teacher school.

Nobody says it out loud, but everyone knows it. It's a whole lot harder to teach freshmen and sophomores than seniors. More experienced teachers want calculus and pre-calculus classes because the kids are easier to work with.

Unless you get the discipline figured out, your class might seem out of control.

The students are older, and they want to pass; they will work quietly and ask reasonable questions. Sometimes, they even express interest. Veteran math teachers feel their dues are fully paid, they are entitled to the advanced classes filled with "good kids."

Struggling math students are younger, they are usually less mature, and they just plain mess around more than older students. And they are yours. If you can't get these guys under control in your classroom, nothing else in this book matters.

As a single forty-year-old when I started my principal internship, I had a lot of time on my hands and decided to study student age versus discipline referrals and cross-correlate that with school subject area. I even subdivided the data by teacher. This was before Excel was available, so the project consumed a lot of time. What did I discover?

More than 50 percent of all discipline referrals went to freshmen boys. Surprisingly, about 25 percent went to sophomore girls. Looking more closely at the freshmen boys, we saw that a large majority of their referrals came after lunch. That particular year, no senior girls were written up for discipline, although they did plenty of skipping. I should add that the orchestra teacher never wrote a discipline referral, and none of his students ever received a discipline referral from any teacher. I am still out there subbing, and I can verify that those numbers haven't changed much in the last thirty-five years. I guess if you want quiet, compliant students, you should teach orchestra—band not so much. As the Pareto Principle predicts, 80 percent of your problems will come from just 20 percent of the population.

The Pareto Principle is a basic tool for analyzing system success and failure. To simplify: 80% of your success will come from 20% or your efforts. Conversely 80% of your problems will derive from just 20% of your students. If you play with this a bit, the math will show you that 64% (.8 x .8) of your problems come from 4% of your students (.2 x .2). And so on.... Most of the 4% will be enrolled in your math class. Enjoy!

When stuck in the loud chaos, your young Pre-algebra students bring with them, it's hard to remember that all behavior, zany or not, is need fulfilling. Kids do what they do to get what they want or avoid what they don't want. It is not helpful to dwell on the idea that the students should come to you with courtesy and respectful attitudes as a "built-in." Many of these kids do not possess what you and I call a rudimentary set of social skills. Now, you can sit in the staff room bitching about their behavior, or you can fix it—or both, I guess, depending on the day.

None of your students will ever be fired from a job because they failed to factor a quadratic. They might be fired if they can't get along with others, refuse to listen to instructions, won't work well in a group, and don't accept responsibility for themselves and their work. The social/emotional content of your teaching is far more important than the content of your math class. Statistically, failure to get along with coworkers is one of the main reasons people get fired. Attendance is up there too.

Classroom management matters. Simple predictable routines in classroom management matter quite a lot. Every unstructured moment in your classroom is an invitation for struggling students to engage in loud, disruptive, off-task behavior. We have talked about building relationships and connections using interviews and other games. We included the notion that you are their friendly teacher—not their friend. We talked about ensuring that every day is a working day with a graded assignment due every day. All these ideas help create a positive classroom. But you will need more.

Once you accept that teaching discipline is part of your job, life gets easier. Think clearly about what your kids should do in your classroom and create lesson plans for a successful classroom behavior. In addition to all the success strategies, you need a comprehensive plan to create an orderly, disciplined learning environment. You will need lesson plans for courtesy, respect, and work ethic. You even need a lesson plan covering how students should enter the classroom when they are late.

At the alternative school, we even had a lesson titled, "How to be successfully arrested."

Start with a theme. You could use any theme, but I always liked the idea of helping my students become "highly employable young adults." Hard to argue with that goal. I had a lot of rules, but most of them are phrased as things to do: the positives. You will have to include some don'ts, but even then, you can begin the sentence with the direction and then include the prohibition. For example:

Use professional office workplace language. Do not use demeaning language.

Honor diversity. Do not make comments about people's race, sexuality, or social circumstances.

Use a conversational "inside the office" voice. Do not swear or curse.

The list of rules in the appendix may seem too long. I found that a few of my students want to litigate every little rule. (For example: "I didn't throw that ruler, I just lobbed it" begets the rule: "Do not cause objects to become airborne—0 trajectory") For me, a good list includes both general and specific statements.

In addition to your own rules, it is a good idea to thoroughly know the contents of the Students Rights and Responsibilities Handbook. It helps in conversations such as this: "Yes, you do have to sit where I ask you to sit. Would you like to look at the law as listed on page seven?"

When you connect with the kids, and they start having some success, your Pre-Algebra Class will look and feel quite different.

The first minute of the first day of class sets the tone for discipline for the rest of the year. Use it wisely. On the first day of a new class, I meet the students at the door. I greet them with a handshake (pre-covid) and welcome them to **my** classroom. You can say this out loud, or it can be the built-in understanding that is the foundation for all your work. I have been doing this for fifty years and observed hundreds of teachers (maybe more than a thousand). Those that start with the notion of "My Classroom" have much better outcomes than teachers who lack this clear vision.

*Let's be clear. I am the teacher; this is **my** classroom. It is not **our** classroom. This is not some sort of a democracy. This is the place of business wherein we go about the serious and important job of learning. I am your friendly teacher (supervisor). You students are the workers (and eventually the clients). The products we produce here, in this professional learning office, are knowledge, understanding, and skill. We owe each other our best efforts in this endeavor. Your parents and society expect us to make good use of the resources and time they have provided. We will do that!*

This clear, direct statement doesn't mean that you need to be a tyrant, a bully, or a mean boss. It means the opposite. These kids are important people, they are valuable, important workers under your direction. As their supervisor, an important part of your job is to teach them the social skills that will help them become valuable group members and thus highly employable young adults. You need to introduce and explain your rules just as you would any important learning activity in your classroom. It needs to be a whole unit of instruction even if you teach it in pieces. You need to check for understanding—a quiz or test. You need to keep records of the test on file—with their signature. Students should sign and date the rules test. You should communicate the rules to parents. (I suggest a brief email with the list attached. They probably won't read it, but it will simplify later meetings if they become necessary.) You should live the rules every day both by actions and words. You should revisit them now and then.

Let's review:

- Accept that teaching appropriate behavior is an important part of your job. Embrace it.
- Develop a clear purpose and philosophy to drive your behavior/discipline system.
- Demonstrate daily that you are clearly in charge of your space. You are the friendly teacher—not the friend.
- Articulate your expectations with a clear list of rules.
- Actively teach the behaviors you expect.
- Communicate the rules and expectations as part of a behavior learning plan.

That's the foundation of a behavior/discipline plan. Here are some important specific actions to help you implement your plan.

Now take a break. Go crack another beer, then tackle this section on the specifics of behavior discipline. It's a long section, but every part matters.

Call parents before there is a problem and try to say something nice. When you are working with first quartile students, the "honeymoon" window can be pretty darn short. I know that you don't have time to call all one hundred fifty of your parents, but do make those fifty phone calls to your struggling students in the first week of the semester. Short, friendly calls work well. You can confirm there shouldn't be any homework if their student works hard in class. Tell them that every day is a graded day, and you expect the students will be more successful with this approach. The parents will welcome your non-homework policy. They have been fighting with these kids about homework for years. You can also remind them to look at the email you sent and invite them to your parent night. With a little practice, you can make between five and twelve calls in an evening so you should be done in about four or five days. Time well invested.

Never, under any circumstances, lose your temper. The minute you get loud, aggressive, or sarcastic the student(s) are in control. If you feel yourself getting angry, do whatever you need to do to calm down and focus on the kid's behavior instead of your anger. You may need someone to cover your class for a few minutes. You may need to have an administrator come in. Remind yourself that this moment at least it's only a job, you are in charge. Get centered, this moment will pass.

One time when I was principal at the alternative school, I got so angry at a kid that I couldn't dial 911. He refused eleven requests to either change seats or go to the office. He told the teacher and then me to insert as many F words as you like. After four attempts and six deep breaths, I finally remembered to dial 9-911—outside line, then dial the number. I thought to myself, "Jim, be careful, you are way too pissed off." More deep breaths, and I started writing my narrative to hand give the police when they arrived. By that time, the other kids and the teacher were sitting in the commons writing up their observations.

The lesson I learned is to be careful. Kids can get to you, and some of them are experts. This kid knew my personal button is chin up, direct eye contact defiance, and he darn near pushed it.

More helpful tips.

Be specific when you address students. Say, "Jack, would you focus your attention on me." Use a name. Avoid general statements such as, "I need you guys to be quiet." With students, "you guys" does not register as "me."

Make positive requests instead of admonitions. If you say, "Velma, please direct your attention to me." You might avoid the argument that starts out, "I wasn't talking." You will hear something like that the first few times that you ask for attention, but you can easily deal with this. In a calm voice respond, "Velma, I didn't say you were talking. I asked you to direct your attention to me." Velma may want to continue explaining that she was not the one talking. Just respond, "Good, great! We agree. Thanks for focusing on me."

Depending on your relationship with the class, you can vary this routine with, "Right now, I am not feeling very special, Levi, and you know I like to feel special when I am helping you learn something." Another helpful variation can sound something like, "Kiki, would you make sure everyone in your group is paying attention? Thanks."

Use procedures to reduce opportunities to mess around.

When materials need to be collected or passed around, we procedures like an elementary classroom; the kids know how to send one person to the supply table—quietly—with no fuss. There is a procedure for clean-up and one for turning in work.

My kids don't raise their hands and ask to go to the restroom. They get up quietly, sign the "out-of-room" sheet, and pick up the pass lanyard off the hook by the door. Do they abuse it? A few do take advantage. However, most kids respect the rule. Besides it doesn't matter. If you want to end your career quickly, just deny a student the right to use the restroom. If they have irritable bowel syndrome or the onset of menses, you are in deep trouble. We could even be talking civil lawsuit.

Be obsessively fair and consistent. Teenagers expect mercy, but they are not about to provide any for you. You must enforce the rules the same way all the time, every time. The tiniest hint of special treatment will lead to arguments of, "What about . . ." If you do deviate from your standards, be sure you have a reason that makes sense when you are explaining it to the principal with an angry parent sitting in the office.

Be sure you're right. If there is the slightest chance you made a mistake, cut the kid a break. If she/he is really a bad actor, they will continue to behave badly in the coming days. You will have plenty of chances to deal with it then. Failure to be fair creates lasting enemies.

Back in the day, the 1950s and 1960s, a time before student rights and eager lawyers, I collected many a hack on my butt. I am guessing more than ten, less than twenty. Every male teacher owned a collection of paddles often made by students in the woodshop—it was a different time. Those teachers left bruises.

Fearing more punishment at home, both my brother and I never told our folks of our misdeeds and the sometimes overzealous, almost brutal, punishment. We knew mom and dad would side with the teachers and punish us more.

I only mention all this because the only one of those "swats" I remember in any detail is the one delivered by the crusty old Ag teacher, RJ. RJ decided I whistled when I walked past his classroom. I was outdoors. I didn't whistle, it was another guy. And why should it matter anyway? RJ did not listen, instead, he lifted me off the ground with a four-foot-long piece of 2x4. My butt hurt like hell for two days. It hurt a little bit for a whole week.

"So, are you going to change your behavior?" "No, but I am going to get even." If a little punishment doesn't work, more is not likely to work better. Kids hate unfairness most of all.

Later, I became a teacher in my old high school. I told RJ off. He didn't care anymore then than he cared ten years earlier. He didn't even bother to shrug. It was a different century—literally.

Don't talk if students are not listening. You can spot a new teacher in seconds. They continue to plod through their lesson even when almost nobody is paying attention.

Of course, they forge ahead. They worked hard on that lesson. They have a good intro. These are important concepts. These kids just need to focus.

A hard truth for new teachers to learn: Your kids did not come to you with respectful silence as a core behavior, and mostly, they care little, tending toward nothing, about your lesson. Instead of talking over students or constant scolding, there are some techniques that tend to work better.

Try changing your position, get away from the front and center of the room. Stand by the "chit-chat" table and continue teaching. If this proximity intervention doesn't work, change tactics.

You can go silent, that's an easy one. Sit quietly and wait. It can take an excruciatingly long time, but eventually, all but one or two kids will stop the chatter and look at you. Directing your "teacher stare" at the last offenders will usually restore order. I should add that, while subbing, I had a particularly notorious class of seventh graders who set the record for silence—twenty-six minutes just to take roll. I think I could fix those guys in two or three weeks. Maybe!

If silence doesn't work, you can escalate to, "Okay, obviously you guys don't need my help. This assignment is due at the end of the period." Then go sit at your desk and grade papers. And this is where the value of every day is a graded day kicks in. They need you, they need those five points. Within a few minutes, a student delegate appointed by nods and whispers will ask you to return to the front of the room and explain. Some of the students will still talk a bit, but you can usually control that with proximity, your teacher stare, and more silence. You may have to use this procedure more than once or twice before they get it.

Take charge of who sits where. In chapter six, we talked about seating. Seating plays an important role in discipline/behavior as well as instruction. Left to their own choices in a classroom filled with rows of desks, students will select seats based on their plans for success in the class. The *A* students will cluster in the first row or two and in the middle row. The kids with moderate interest, *B* and high *C* will sit right behind or next to them. The "chowder heads" will collect in the back corners. The hard-core non-achievers will favor the left corner of the room as you look at them from the front. They instinctively know that teachers tend to look more often to their right than their left. You can use this knowledge to improve behavior/discipline.

Back in the day, I tried a couple of experiments with my all-male, mostly difficult, freshmen classes. I had a small classroom with five rows of tables. On the first day of class, I greeted kids as they came in and let them select their own seats. After everyone settled down, I moved all the front row people to the back, and the back row came up front. Same plan for the second and fourth row. The middle row of the five rows stayed in their same seats. That worked fairly well.

After a year, the kids heard stories from their brothers and kind of caught on (we changed classes four times a year). But as they walked in the door, it was obvious which kids were front row and which were back row material. I just didn't move anybody. One young man even asked, "Aren't you going to move us?"

"Nah, you're good." And they pretty much were good.

Here is another approach to seating: Leave the classroom in rows for a week or so. After you establish in your mind who are "first-row kids" and who are "back-corner kids." You can assign table groups. Over the years, I have developed a little system that works well. I use table groups of four. I let kids "request" one person (and one alternate) that they want in their group. I try to assign seats so that the two buddies, pals, friends, whatever they call it, sit together. With practice, you can usually put one "chowder head" at each table along with a friend who is only a semi-chowder head along with two front-of-the-room types. This gives you a little extra leverage on the off-task-fool-around guys. You can always threaten to move one of the buddies if they don't regain focus.

Document early and often. If you want support later, you need a paper trail. Your principal needs to demonstrate that you used a progressive approach to discipline. Hopefully, by the time you need the principal's help, you have documented that you sent home a letter on behavior expectations, and you had the students sign a copy that you keep on file. If you made the "good" phone calls at the beginning of the semester, you should have kept a record—even it is only a checkmark on a class roster.

If you are having some struggles with a kid, get on the phone with parents. Do it early. Don't wait until it's a big problem. Be friendly and gracious. If you make the call during your prep period, you can call the student out of class. Your call will have triple impact if the student is sitting near you. Plus, you take charge of the communication before the student contacts home with a spruced-up version of the story. You control the narrative. The students will be less likely to minimize and redirect if you are sitting next to them.

I know you're tired and not looking for an unpleasant parent call, but sooner is way better than later.

"Hi, this is Jim Slosson, Matt's Math teacher."

A typical response, "What's he done now?"

"Oh, it's not all that a big deal, but I thought it would be best if we could all get on the same page early in the semester. He's a pretty great young man, but that darn phone is getting in the way of learning. Do you see any of that at home?"

Typical response: "Yeah, he never gets off that damned thing."

"Well, I am hoping that I don't need to get the assistant principal involved, and we can take care of it between us. Do you have any ideas?" Before you finish, offer the student a chance to speak. Usually, the students don't have much to say.

Sometimes the parents will help you, sometimes they won't. Sometimes parents can't help because they don't have much control over the kid anyway. Make a short memo for record about the call and ask the office to enter it into the kid's discipline file. Write something on the order of, "For record only. No action requested or needed at this time."

Back in my principal days, I ran into this discipline problem often: teacher is fed up with kid. Sends kid to office. Teacher wants kid kicked out of class. "How long has this been going on?" I ask.

"For weeks!" the teacher replies.

"Have you written him up? Have you contacted parents? Have you made some anecdotal notes?"

"Jim, you know I don't have time for all that." (Sometimes they used to tell me that I don't know what it's like in the classroom, and I would remind them that I spent twenty years working with the toughest kids before I was an administrator. I paid far more dues than they ever will. If the teacher really irritated me, I would tell them that if they don't like kids more than English literature they are probably in the wrong business. But I digress.)

"Well, that's unfortunate that you didn't have the time. Since he did not do what the law calls exceptional misconduct, all I can do is assign a low-level corrective and call the parents." The teacher then goes to the staff room complaining that there is no backup or support around here. The principal complains about the lazy teacher to other administrators. The kid continues to be a jerk.

When I was a union grievance rep, we used to say, "If you did not see and hear it with your own eyes and ears, it didn't happen. If you didn't write it down, it still didn't happen."

The trick then is to start taking notes of some sort as soon as you think there might be an issue in the making. It can be simple. I still use a paper roll book every day. If I have a discussion with a youngster about behavior, I just put a little mark on the grade book—a little dot.

If a kid collects three or four of these little marks, I write an email to the student with and "cc" the parents and assistant principal, and maybe even the counselor. It might sound something like this:

"Isaiah, on four separate occasions, documented in my roll book, I have asked you to stay in your group and keep working. In every case, you were slow to respond, and two times, you ignored my requests. I know you heard me because you made eye contact. If this behavior continues, I will initiate progressive discipline steps so you can correct your behavior and demonstrate the habits of highly employable young adults."

If young Isaiah still doesn't get it, I start writing him up formally on the next infraction. Usually, the kid comes around. Importantly, Isaiah tells his friends, and they start thinking that they might want to avoid such an email with a copy to their parents.

What to do when you get ambushed. After forty-nine years of teaching, I still made a rookie mistake. I was called to the principal's office to visit with a complaining parent and student, and I screwed up. The principal said, "Would you mind telling us what happened from your perspective." I started talking. I should have waited and said, "Well, since you folks have already discussed this without me in the room, why don't we have Josias repeat his story, then I will be caught up." It turned out fine, but I was off balance, and we had a bit of an argument back and forth. I commended the parent for advocating for his son, but I also pointed out that sometimes, teenagers leave a few things out of the story. If you get a surprise meeting, don't get in a hurry to defend yourself. Take some notes, then reply.

Sometimes, a little punishment goes a long way. Here are two little things you can try. For minor transgressions, I sometimes use LL, last to leave. It just means that the student must wait and leave the room last when the period is over. One tiny notch higher on the progressive discipline scale is LTA, lunch time annoyance. Suppose a kid deserves an email, but you offer the student a choice of LTA. Instead of lunch detention, the kid just shows up in your classroom before he/she goes to lunch. You only need to detain the student for a minute or two. It's annoying but not really a punishment of any consequence. They get to lunch late, go to the end of the line, get stuff they don't really want, no big problem. What if annoyance doesn't work? Up the ante.

"Sigh, can I go now?" "Yup. Thanks for dropping by."

When you have some leverage, and you really should do a formal discipline referral, you can offer the student the option of lunch in your room. "So Billie, I can turn in this write-up, or you and a friend can come in here and have lunch with me." Some kids will take you up on the offer since it is less unpleasant than a day sitting in the detention room that some schools call the restorative behavior lab—or some such thing—it's still detention.

Schedule a couple of kids in at a time. Bring them a treat. Ask about their pets, their cars, their sports, their plans for the weekend. It doesn't matter. Just let them talk. Ignore the behavior issue. You and the students will connect, they know you value them. Their behavior will improve. Warning! They may start to enjoy the time with you, and you might end up with a lot of lonely kids eating lunch in your classroom. Not a bad problem. This time is about conversations not about devices. Just make sure these youngsters are interacting with you and not with their phones.

Even for me, this was pretty far out there! I had some of my alternative guys in for Saturday school and told them to bring quarters and meet me at the driving range. We spent half an hour hitting golf balls, and I sent them home. Their behavior improved, and we met at the range a few more times even though they weren't in trouble. My assistant thought I was crazy, "What kind of punishment is that?" The punishment was getting up at seven-thirty on a Saturday. The golf was a connection for a better outcome next time.

Ah, phones—Lord have mercy

When I sub, I love cell phones. I hand out the assignment. The kids mess around on their devices. They play games, text their friends, and watch videos. They don't cause any trouble.

When I am the regular teacher, cell phones are hell. Some kids feel my lesson is an interruption to their busy, virtual day.

True story. Picture a young man sitting in the principal's office (my office) for his final interview before I sign off on the paperwork, and he is truly hired as a teacher. Button-down shirt, ironed khakis, polished shoes, crisp haircut, neat mustache. I am in mid-sentence administering the "blood oath" * *when his phone vibrates. He pulls it out of his pocket, unlocks it and checks his email/text—whatever. He spent five or six seconds looking at the screen and then looked up. I asked, "Have you got something more important to do than this interview?"*

He replied with an "Oh, sorry." But he really wasn't sorry. He wasn't even embarrassed.

I remained silent and looked at him. Finally, he responded, "I'm not getting this job, am I?"

"No," I said, "you will not be working here. But you have a good paper, and you interview well. Something else will come along. If I were you, I would leave the phone in my car for my next interview."

(*The "blood oath" will be explained in the second book of this trilogy titled *How to Fix a Broken School.*)

Devices are an addiction. How do you want to deal with this addiction? Some schools try to ban phones completely. That can kind of work—sometimes—even a heroin user can go a few hours without a fix. But if you go that route, you and your administrators are going to pour a ton of time and effort into that fight. Here is how I have made my peace with devices. It mostly works.

I don't give kids an entry activity. I take roll the old-fashioned way. I check off names on a paper roll sheet—make eye contact with the student—say something nice about each one. The kids get in their last, vital, (sarcasm) social posts. When I have entered attendance into the computer, I ask them to put the phones face down—mostly they do that. I start my lesson/explanation/demo, and they go to work on their assignment. Now it is time for forty or forty-five minutes of actual graded work on the non-homework assignment. I walk around the room helping, monitoring, directing, coordinating as they help each other. If a student has finished the assignment and it looks good, I mark 5 on the top of the paper, then the student may use a phone. If the assignment is not done, and the student is using a device, I often say nothing. I just use my white board pen to make a dot on their paper. Sometimes I might add, "It would be swell if you got back to work."

When I enter the grades for the day's work, I check for dots. A dot earns a minus point in the "device" column. A couple of interesting things happen when you add in a negative point. Kids question me, "Why did I get minus points?"

I explain, "Because you were on your device when you were supposed to be working. When you get five negative points, I will contact you and your parents by email, and I may even follow up with a phone call."

Who knew? I must confess to you that I didn't realize you could give negative points on an assignment until I was dinking around with the grading program during my forty-sixth year of teaching. We use Skyward. I have no idea if it works on other systems.

What if a kid finishes way too early with a complete and correct paper? I stick with my bargain. It will even out in the end, and it lets kids know that I keep my word even when I don't like it. Complete work done well should earn a reward. Some kids who finish early start working on stuff for other classes—rare, but it does happen. Much more likely in Algebra II than in Pre-algebra, Algebra I or Geometry.

Pencils—really! Do you have time and energy to shed blood on the pencil hill? I like to keep enough sharp pencils on hand so we can avoid pencil sharpening while I'm talking—in fact, it's a rule.

Don't argue—don't give reasons. For all kinds of purposes, and in all kinds of ways, kids will want to engage you in arguments about behavior. They will ask for explanations, reasons, or proof. Rookie teachers try to explain. It never goes well.

Veteran teachers know that explaining just leads to more arguments—more fodder for your young, would-be attorneys. The more you explain, the more they will litigate. For example: "Why do I have to move? I wasn't talking" or alternatively, "Everyone was talking, why do I have to move?" You don't need to argue. You can end debate by saying, "My instructions are clear, please follow them." Then walk away. The kids will usually comply, but if they don't, you may need a longer response.

Suggested response: "I did not say you were talking, I asked you to move. I may or may not have reasons, but you do not need to know them. You only need to know that you must move because I wish it to be so." Or use whatever version of that logic that suits your style.

If it continues, you can move up to: "I have asked you to do something that is reasonable and legal. It is something that I think will improve the educational process. Are you telling me that you are refusing to follow a lawful order as outlined in the student handbook and state law?" A kid must be damaged, obnoxious, and kind of not too smart to go beyond that reply.

But it does happen. If things keep going south, you will need to write the kid a discipline referral.

Do your first write-up well, and it might be the last one you need. You must write in clear, factual, and honest prose. Include highly specific details. Use direct quotes.

For example, I used a clear monotone, almost robotic, voice and repeated for the third time, "Please return to your seat." You turned around quickly and stepped toward me. You raised your voice and tilted your head and chin upward at about a fifteen-degree angle. Your body leaned forward. You made direct eye contact and used a challenging, confrontive voice. You said, "Get the fuck off me. I will sit down when I'm fucking ready." I stood in one spot, about twenty-four inches away from your body, and you pushed me on the chest at sternum height. You used your right hand. Your fingers were splayed. You did not make a fist. You pushed hard. I did not move. You said, "Get the fuck out of my face."

At that point, Mr. Furgeson arrived and directed you to go with him. You turned left and started to walk away. As you reached the door, you looked back at me and said, "Fuck you, Slosson! This isn't over." Your tone of voice and body movements were confrontive and threatening during this incident. I felt that if Mr. Furgeson had not arrived, you would have continued to assault me.

Your write-ups will probably be less dramatic, but they should be just as specific and detailed.. The sooner you write it all down, the better. If you have a super difficult incident like mine, ask someone to cover your class. You need to calm down anyway. Go to a quiet place and do the write-up immediately. It will help your principal, and maybe even the police. Don't forget to list witnesses.

Important. The write-up should not include any speculation about what is driving the behavior. Try to keep it non-judgmental. Instead of saying "stole" or "theft," use terms such as "possessed property that was not hers." Avoid trigger words.

When I worked in Sumner, Washington, most parents worked in factories. If I called and said, "Mason skipped today." The parent would say something like, "Dammit. Give him Saturday school." If I said, "Mason was truant," the parent would go crazy and threaten a lawsuit. Learn the trigger words for your parents.

Now, before you get freaked out, you should know that in fifty years of teaching, I have only had to write a letter about assaultive behavior on my three times. And remember, I have always worked with the most difficult kids you could meet in a public school. You probably won't have to deal with anything that severe ever.

But if ever a kid does lay hands on you, make sure that it gets reported as an assault. If your principal refuses to make the call, do it yourself. There must be some behaviors that transcend the bounds of

PBIS and "restorative practices." If they fire you, get a job in another district and run for school board.

And sometimes, there just isn't much you can do. Even if you do everything I suggest, there will come a time or two in your career when you have a discipline situation that even you, with all your magic, just can't fix. I've had a few. Let me share.

The next two incidents I will describe are not factually true, BUT they are composites of many incidents from my alternative school days and clearly illustrate, that you, the teacher, can't always fix everything in your own classroom.

The counselor put Linda in my class in late October because she was "misplaced" in Algebra I, which really meant she as a really, really, difficult young woman, and she walked all over the rookie teacher. Normally, I could have handled Linda. But Linda liked to taunt another student. Linda was having sex with the father of Kaitlyn's baby. And Linda liked to provide Kaitlyn descriptions of their deed-doing.

Anyway, I went back to the counselor who said, "Well, I don't have any place else to put her." I wanted to say, "Let's just make her make her your TA." I didn't say that or any other smart aleck stuff that came to mind such as, "Are you an actual counselor or just a schedule-filler-outer?"

I did say, "Well, Linda has got to go, and you need to help clean up this mess. Let's go talk to Gene (the principal)." Luckily, I had four of my more exquisite discipline referrals in hand. He had plenty to work with in dealing with the young ladies. I explained the situation and dropped off the referrals. When I left the office, I could hear Gene talking to the counselor, "What were you thinking?" The implication was clear: "What the hell were you thinking?"

The second case was similar but less personal. Matt was a low-level gangster, and I had him all socialized-up and working on math and being a responsible kind of guy, if only in class. I had invested a lot of time with his aunt. Both his parents were incarcerated. Angry kid. Natural leader. Wanted to be the alpha male in the room. It took time, but I was in charge. On the first day of March, in comes Max, another wanna-be alpha gangster. They bellied up to each other that very day.

They called each other out the second day. Fists flew on the third day. Security removed both of the guys. They were emergency expelled.

I took care and finished my write-up with extra attention to each detail. Thank the Lord Matt remembered to take the first two punches before he hit back. Max went back to the incarceration from which he had come. Matt got a five-day suspension. His probation officer delivered a packet of assignments to juvenile detention.

Now if you are a rookie, nothing in your teacher training or life experience has prepared you to deal with these kinds of discipline issues. You need help. Do not be embarrassed to seek it out. Not all kids can be fixed in your classroom. But don't use the Maxes and Lindas of the world as an excuse to give up trying to fix the kids you can fix.

Chapter 14

A hard copy for CYA, and send some email. Twice a month is enough—and keep them short

Your first parent communication should be a short, hard copy letter about the class that includes your school email and school phone number. Attach a set of rules. Require the kids to bring back both the letter and the rules signed by a parent. If it's a 10-point assignment, students will bring it back. If they don't return it, put a -10 points in that grade book slot. The student will bring it back soon enough. You need to keep these signed papers in a special drawer so you can find them easily.

This is just pure CYA. You only need the papers if the student or parent claims they didn't know the rules. You will find, in the rare event of an ugly parent meeting, that we can more quickly move to discussing student behavior instead of teacher communication skills. What if the kid forged the parent signature? Not a problem; that's on the kid, not you. In fact, you can change the conversation to a discussion of the kid's devious behavior in addition to whatever happened in class.

Back in the Alternative School. I meted out some trivial punishment for the first offense of smoking, most often crushing enough aluminum cans to fill a garbage can. Usually, the kid would say something like, "Aren't you going to give a warning for the first offense?"

I would reply, "I respect you too much to give you a warning. You signed the rules when you came to this school. That was your warning. If I warned you again, it would be like saying, 'You're not very smart.' I think you are smart, and I am not willing to disrespect you in that way." What is the kid going to say? "No, I'm dumb, give me a warning, please." Well, one did try that approach. I smiled and pointed to the door.

Parent Meeting. In addition to the letters, offer the parents a group meeting in your classroom during the evening. Try to do it the first or second week of school right after you have made the "good" phone calls. Serve some punch, water, and cookies. Don't overbuy, not many parents will show up.

Show the parents the results of their students' initial assessments. Explain the instructional materials for the course and explain your teaching methods. They will sigh with relief when you tell them there should not be homework. Even if you only get a few parents, this meeting is worth the trouble. If your evaluator is working in the building that evening, invite that person to drop in. They might even show up, administrators work punishing hours so there is a good chance they will still be working at 7:00 p.m. Everyone will love you more if you keep the meeting short and end on a positive note, "We're going to have a great year!"

General emails. We have it easy. Back in the bad old days if teachers wanted to send emails to parents, we had to build our own databases. Today, you just go to the message center on your grading program, and you can send emails to all students and parents by clicking a couple of boxes. Use this feature, but don't overdo it. If a little is good, more is not necessarily better. If you send too many or if they are too long, nobody will read them.

Keep the communication short. Generally, parents don't scroll down so say everything in one or two tight paragraphs. Always stress the positive. "Students have completed two tests and twenty assignments. If you see a zero on a test that will dramatically reduce the grade, encourage your student to make up that test soon."

I send reminders to students, cc to parents, reminding them about tests, due dates, and catch-up days.

Report cards are another chance to communicate with parents. If you really dislike the kid, leave the "enjoy having student in class" comment box blank. Otherwise, you like all your students and really enjoy them so check the box. Be careful with the negative comments. You only need to check one or maybe two at most. If the kid is lazy, mouthy, or frequently tardy, the parents already know it.

Your school probably has some kind "Spartan, Owl, Viking" whatever pride cards. Fill those out for kids who are making a real effort. Try to complete two per class per week. Here is the kicker, put them in an envelope, and mail them home. Find a TA with good handwriting to address the envelopes, OR have the kids address the envelopes themselves, you will be shocked when you find out how many kids don't know their own address.

Huge positive impact when the parent opens the mail and warm fuzzy falls out. Warning! You might want to write, "Good News!" on the outside of the envelope. Your students' parents aren't used to pleasant news from the school.

Chapter 15

Implementation

Cut yourself a break, you don't have to do everything at once.

Even if you devote your entire summer to math, you can't do everything I suggest in one year. When I began this journey, I had just retired as a principal from an alternative school. I gave the project sixty hours per week, and that was just to develop the lab projects for the class. At the time I planned to use the old Saxon light blue book, Pre-algebra.

I started teaching my success strategies class at Elton High School in September. From the very first day, the labs worked well. The book not so much. The kids just wouldn't do the problems from the book. Of course, they wouldn't.

One evening, I was helping my fifteen-year-old son with his algebra, and we were both a little stuck. I said, "Here, give me a minute to read the text, I can't remember how to do this."

Bill was stunned, "You mean you can learn math from the book?"

"Yeah, what do you think a math book is? Just a package of homework problems?"

Bill's answer, "Yeah, pretty much, I guess. I never really looked at a math book very carefully." And neither do most kids.

After a frustrating first month at Elton with no progress using the Saxon book, and a couple of chats with my wife, Pat, I decided "what the hell," I will just take the Saxon pages and rewrite them as handouts: One sheet of paper both sides, handwritten, due at the end of the period.

Unbelievable!

The class took off! Almost every kid was working the whole period. Even the two lads that slashed my tires focused on math for the entire ninety-minute period.

Within in a week, I started to add little embellishments of my own to the handouts. It worked even better when I decided that kids who finished early could play chess, checkers, or gin rummy. (Phones were not a big deal in those days.) They still groused about the grading system since they believed it was their God-given right to do lousy work, but they worked hard every period, every day.

The handwritten assignments worked so well that I spent every weekend writing out three or four assignments until I had ninety assignments and ninety labs. The second year proved to be way easier than the first, and I rewrote all the assignments, so they matched up with and reinforced the lab projects. I included a quite a bit of eighth grade geometry. I addressed some of the antecedent learnings that the kids lacked such as multiplying and adding common fractions. We included multiplication facts, and math vocabulary. We also worked on oblique and isometric drawings and folding paper to make 3-D shapes.

Four years and four complete rewrites into the project, my computational assignments didn't look anything at all like the Saxon book. About the only thing that remained was the Saxon idea of rolling review. Each assignment included a few problems from earlier in the year. I decided it was safe to copyright the material as my own.

I doubt that you are going to have the kind of time that I had available. I was at that place in my life where there were no kids a home, and no *parents to* care for. Helpfully, Pat was just as obsessed with her work as I was with mine. And I didn't have a boat.

As I write this book, I am thinking of how I might provide you will all the Pre-algebra materials and labs that comprise the second edition of Math Lab also known by its Elton name of AGA – Algebra-Geometry-Arithmetic. Since 2006, I have never let released my stuff unless the teachers attended my in-service training. I'm reluctant to change that policy, but maybe it's time. All those materials aren't helping anyone just sitting in a box in the garage, and I'm pretty sure my kids will just take them to the dump when they park me in a place where the Hallmark channel plays 24–7. (I did change my mind. There is a how to order page at the end of the book.)

You need to implement the success strategies in a way that fits into your work/lifestyle. Begin a bit at a time, start with things easy to implement. Treat yourself like your students: let success create more success.

Maybe this list will help. The order suggests easiest changes with highest payoff to get you started.

Interviews and name games. You can look at the interview directions in Appendix B and easily implement this idea. Make up some cards and give it a try. As Tina discovered, this one activity changes the whole atmosphere of your classroom. Struggling students will enjoy your class more, and you will find they work with a bit more enthusiasm. At the same time, use some of the name game/strategies to ensure that all the students know the names of almost all the other students.

Arrange your room to allow students to work in groups. This change takes little time and no money. This high leverage* strategy helps you in two ways. If you can figure out a class procedure to get the backpacks off the floor, you can wander around and monitor kids' work. And since the students now know other students' names, you can ask them to help each other. The kids really like the "free problem" technique. I wander from group to group and say, "I'll do a free problem with you guys, which one do you choose?" After a bit of give and take, they would pick a problem, and I would lead them through it step by step which of course they had to rewrite to comply with the "showing all your work all the time" requirement.

*High leverage changes bring large amounts of benefits with low expenditure of effort. For example, doing interviews and moving chairs will create achievement gains for free. Low leverage changes require great effort for small return. Buying a new math program fits nicely into the low-leverage category.

Change your grading system to points—get rid of weighted categories. This change costs nothing, and it improves communication with parents and students immediately. Since you're a math teacher, it won't be a big problem to figure out how to make tests 80 percent of the grade, and everything else 20 percent of the grade. In one high school, I tried to change my system away from weighted averages and the district would not let me. After several conversations, they did not say, but I came to understand that they would not unlock the grading system on the computers because they didn't trust the teachers. They feared someone might not keep the 20 percent daily work and 80 percent test grade ratio they wanted. Funny! See "work-around" in the grading chapter.

Start using practice tests as part of your test review. If you tell kids that they will get two chances to take the test, most of them will just blow off the first test. What if you could change that up? What if you could get them to do their best on the practice test?

You don't really need anybody's permission for this little change, and it doesn't take much effort—especially if you handwrite the tests yourself. I generally use the same questions on both tests. I just change the numbers in the problems.

Creating daily worksheets—non-homework. You can begin using daily worksheets (non-homework) on just one unit. The "every day is a graded day" plan works best when students have a worksheet. I have discovered that, with some practice, I can write about four work sheets every weekend and still have time for family, yard work, sailing. Well, maybe I slack on the yardwork. I used this method for classes in senior test prep as well as Geometry and Algebra. Some samples are included in the appendices.

I dropped by to sub for Pat, one of my former colleagues from Elton. Pat taught science before he taught math, and he was skilled at adapting the success strategies to higher level math classes. Now he teaches four sections of pre-calc and one section of honors/AP stats. His kids work on handwritten worksheets. The students quickly informed me that, "We are allowed to work together and help each other." I said, "That's great." I thought, "Good thing. What the hell is synthetic division?" While they worked, I looked it up on YouTube. The kids get a worksheet every day. Not surprisingly, all of Pat's students score in the top two stanines.

Success strategies work for all kids.

Chapter 16

Your Teaching Schedule—Your Career—
The Payoff is More than Money

I've gone to a lot of trouble to deglamorize teaching struggling students and show you the irreverent realities of working in the system. A few days before the manuscript was scheduled to be sent off to the publisher, I realized that you need to hear about the immense satisfaction of a career in teaching.

Every now and then, sometimes each summer, sometimes a few years apart, three former students, Sandra, Danny, and Greg come to Olympia. We met for breakfast or lunch. We had a few beers or a glass of wine. We talked about careers, grandkids, divorces, people we knew in school, politics, retirement in general, and Medicare. These "kids" are now in their sixties, and the message is the same every year. The four of us mattered in each other's lives. Back in 1976, each of them was struggling in their own way just as I was struggling with my career and my private life. We were creating a yearbook, but more importantly, we were building a support system; our lives were connected, we mattered to each other. We just didn't know that back in 1976.

In addition to our class of '76 meetings, I occasionally have lunch with former students, and it seems every month, I run into some adult who thanks me for being their teacher in high school. "I hated math, but I liked your class."

One of the more interesting lunches started with a Facebook contact. "Do you remember me? Do you remember how you used to talk to us about money and getting rich slowly?"

"How could I not remember you? What other sophomore ever asked about the opportunity cost of money?"

"Well, I did what you told us to do. I am hoping I can buy you lunch and brag about it."

He was a little disappointed because I wanted pizza instead of lunch at an upscale restaurant. After lunch, I checked him out online. Not only is he wealthy; he is generous to his employees and family. He never cut a corner.

I want to leave you with this thought and suggest some ways to make your career experience as satisfying as mine.

Teaching is just about the best thing you can do with your life. In spite of all the nonsense and meddling from the state and the district office, you are in a unique position to make the world a better place. You can't end war, famine, or disease, but you can make the world a tiny bit better one kid at a time. You can't save every student you meet, but in a full career, you can improve the lives of hundreds—maybe even a thousand young people. That is an honor and responsibility most people will never find in their work.

Maybe that's why they don't write country and western songs about teachers.

Some suggestions:

Teach at least one class that is more like a club than a class. Advise the robotics team, do the yearbook, teach newspaper, coach a team, or advise the leadership class. Do something with kids that is fun. For me, it was our vocational industrial club, yearbook, and our video class. You will recruit these kids from your math classes. These are the kids you will know your whole life. You will be invited to their weddings. You will hear from them at Christmas. Sadly, you may even speak at their funerals.

Teach two sections of Pre-algebra. Then loop the kids back in for Algebra I the next year. You will already know their names. They will know you. They will know your system. You can build on the math and the social skills you developed last year. Once you have made those kids your own, your work will be much easier and more satisfying. Maybe even fun. Besides, if you send "your kids" over to a regular Algebra I class taught in a regular way, they will just wither.

Your kids will tell their younger sibs and friends how your class works. You will have a following of young people who already know how you do it. New students will ask, "When do we start the interviews?"

Parents too will support you. They will try to put kids in your class who don't really belong there. Teaching is never easy, but it gets less difficult once you are established.

And one last thought, let's talk money: You won't get rich teaching, but you can do well enough and have a comfortable life. At 10 percent interest, the historical average of the stock market since the Civil War, money doubles every 7.2 years.

If you save just a hundred bucks a month in your 401K or your 403B, it will only cost you $80 out of pocket. When you retire at age sixty-five, you will have about $900,000 in addition to your retirement and social security benefits. If you save half of every raise during your career, you should have a tidy two to three million, maybe more.

That should buy a few nice lunches and some top-shelf beer.

Chapter 17

Preparing for State Tests
"The harder we work, the luckier we get." —Samuel Goldwyn

With minor differences, every state uses some sort of end-of-course test or state test. Our state relies on the SBA* or alternative tests. Many states accept ACT* or SAT* scores as an alternative method to satisfy the graduation requirement. Some states accept ASVAB* scores. Usually in high school, a high school diploma hangs in the balance. Your kids probably need to pass one of these tests to graduate. If you spend a week preparing, your struggling students have a better chance of squeaking across the finish line.

*(SBA—Smarter Balanced Assessment, ACT—American College Test, SAT—Scholastic Aptitude Test, ASVAB—Armed Services Vocational Aptitude Battery)

Since it is the easiest test, let's focus on the ACT. But this procedure will pretty much work for the SAT too. I had to learn this the very hard way. Feel free to benefit from all my trial-and-error screwups.

Do not expect these youngsters to go home and register by themselves. Most of your kids will need a lot of help from you to sign up for the ACT. If you leave it to them, it will not get done! For the most part, their parents will not be able to navigate the sign-up process. Normally, you won't need to do much to get kids signed up the SBA. Somebody else in the district will take care of that.

Test review and practice

Get a copy of the several sample tests available. Plan on working through the whole thing with the class. You will need to budget several days for this review. Your goal is not to reteach your kids the math, but rather to get them familiar with the vocabulary, the test layout, and a time/value strategy for answering questions.

Begin with vocabulary and phrasing. Make sure the students know specific math words such as term, expression, equation, and other words that appear in the sample test. The test uses language such as, "What is the value of the variable term in the following expression?" That wording is not familiar to struggling math students; in their minds the question should be, "How much is x?"

Ensure that kids have the right calculator and know how to use it. The ACT rules call for very specific calculators—no phones. Make sure your students know how to look up the trig functions, and practice finding roots and exponents. Practice more than you think is necessary. They feel a lot of stress when they sit for the test. Stress leads to confusion, frustration, and mistakes. Overlearning helps them remember.

My students work on a printed copy of the sample test, and that approach has been successful. We work together to analyze each question.

- What are they asking you?
- What strategy will you use?
- Can you answer this question?
- Should you move on?
- How are you doing on time?

Time is the enemy of the ACT. Kids have sixty minutes to answer sixty questions. They need about thirty-two correct questions to pass. There is no penalty for guessing. Students cannot afford to waste even a moment on questions they cannot answer. In the words of John Wooden they need to "Be quick, but don't hurry."

Stress over and over during your review that they should look at a question and assess the probability that they can do it. If they can, great, do it. If it looks impossible, teach them to make a quick guess and move on. Don't look back until all sixty questions are answered. If they have time left, rare but it happens, they can work on a few of the "guess" questions.

I like to meet my kids at the test site parking lot about forty-five minutes before the test starts. I bring stuff:

- Extra calculators
- A copy of each student's entry ticket two copies wouldn't hurt
- Sharpened pencils—three for each student
- Doughnuts—or pastry or whatever.
- Coffee and hot chocolate.

Test Day! Plan to get up early. Put the doughnuts in your car. If you want to make up a jug of coffee or hot chocolate, that would be a nice embellishment. You'll need napkins or paper towels and some paper cups.

I like to show up at the test site parking lot an hour before they open the doors. I greet the students and their parents. Invite them to have a treat. I double check to make sure they have a calculator and an entry ticket. I pass out pencils if they need them. I calm them down and express confidence, "You can so do this." I remind them that they only need to get thirty-two questions right, "So don't waste time on the ones you can't do! Above all, don't get whacked out. We have other chances. Show me some mellow intensity. Be quick, but don't hurry."

It depends on my mood and my wife's plan for my day of course, but sometimes, I go to breakfast and wait, then I go back to the test site. I hang out in the parking lot and congratulate the kids for their efforts and collect my stuff. I remind them pass or no pass, I am proud of them.

The following Monday, we debrief, congratulate ourselves, and prepare to wait for the results. And we start discussing the SBA, which is a very different test that requires a different review and test strategy.

The SBA is a bit different.

The SBA is not a timed test. Your review strategy should be a bit different than the preparation to take the ACT or SAT.

Go through the sample tests available. Review vocabulary and phrasing. Teach kids to weed out questions they can't answer. Time is unlimited, but in reality, they can only concentrate for an hour or so. They should bypass the questions that are hopeless.

Kids need a guessing strategy since the scoring includes a penalty for wrong answers. If the student can narrow it down to two answers, they should go ahead and guess. If the zone of guesstimate is three answers, skip the question.

On test day, I get somebody to cover my class and greet the kids as they enter the testing area. I provide candy and sharp pencils, and a pep talk.

I bring a box of graph paper just in case the proctors forgot my request. I can't give it to the kids, but I can give it to the proctors who put it out for student use.

We debrief the next day. Even without word problems, it's a hard test, and I want ideas to better prepare for next year's class.

Acknowledgments

I could not have developed and improved the Success Strategies approach to teaching struggling math students without help and support.

Superintendent Alan Burke had the vision to see that many more kids could succeed in math with a different kind of instruction. He gave me support, resources, and boundaries.

Lois Baker, Director of Teaching and Learning gave me plenty of slack, and lots of logistical support, along with wise guidance.

Jeff Loupas, our stats guy, is a great teacher in his own right, and he provided support, ideas, and a steady hand on the tiller when I needed it.

My buddy, Tony Judah, taught the same class with the same kinds of kids, and he was there every day with suggestions and improvements that we incorporated into the materials.

Lisa Spitzer taught a section of Pre-algebra while I was working on the third or fourth draft of AGA and checked all the math providing corrections and valuable feedback.

Illustrators

K. Murphy is an illustrator and animator who grew up outside of New York. She has worked on a range of projects including logo designs and children's books. She struggled with math as a student and was frequently yelled at for drawing cartoons in her notebooks. You can see more of her work at https://stennyart.weebly.com

Daniel Oviedio is a resident of Bogotá, Columbia. He has commercial experience as a cartoonist, an illustrator, and a graphic designer. Daniel has worked with published authors across the United States. His work can be seen at https://daniel34.jimdofree.com/cartooning-book-illustration. Daniel created the cover Success Strategies and the impish students.

And most importantly

Last but foremost, I must thank my patient wife, Pat Slosson, who put up with years of my obsession in developing Math Lab/AGA and creating this book. It takes a special person to be okay with a guy who gets up at 3:00 a.m. to start work.

Ordering Math Lab Materials

Normally I don't sell the Math Lab/AGA materials unless the users attend my in-service, but my notebooks aren't doing any good sitting in my garage. However, since you read the book, you understand that the magic isn't in the assignments, the magic resides in you and your ability to connect with students. Therefore, the Math Lab/AGA materials are available under the following conditions.

When you purchase the materials, you are buying a license. You own the right to use the materials in your own instruction. You may not forward them to other teachers or other schools. The ownership rights to the materials and limitations are explicitly covered in the packet you receive in the mail.

What do you get for 110 bucks?

- Sixty computational assignments—these are four-page assignments written for ninety-minute period. That means that you have more than enough material for 120 days of lessons. Answer keys are included.
- Twelve tests, two versions each for the computational lessons.
- Eighty-nine lab activities some will need one period of class time, some will require two periods. Answer keys are included.
- Eighty-nine lab reviews of one page each. Print two back-to-back and you have an excellent assignment for substitute days.
- Six pages of reference notes for students to keep in their binders.

- Teacher notes for each lab. Includes tips for teaching lab lessons and games for team building with your kids.
- And a bunch of random stuff such as memory aid ideas like "C and D connect by three and radius looks like half to me" "Area-squaria."

Individual teacher paying out your own pocket—$110. I will send you a flash drive that has more stuff than you can imagine. You must agree that you will not pass the materials on to any other person or entity. If you think it wasn't worth the money, I will grudgingly refund your $110 when you send the flash drive back. You may purchase the materials on the website: prealgebra.net.

District Office $500. I will accept a district purchase order and invoice the sale along with the shipment of the flash drive. You will be granted a district license for up to five individual teachers. Additional licenses are sold in bundles of five for $500. A stipulation of the license is that the materials may not be shared outside of the purchasing district. District purchases are not refundable. The website will provide you with a mailing address for the purchase order, or you may transmit the purchase order digitally to jim@prealgebra.net. Let me know if you are in a one-school district. We can work something out.

District in-service and unlimited license within the district $5,000, plus travel expenses including airfare, lodging, rental car, etc., to be billed on a separate invoice. The district will be responsible for the cost of all printing and binding and the purchase of Meyers-Briggs Inventory materials. In this two-day experience, teachers will receive a detailed in-service manual and experience the success strategies firsthand. Small districts may cooperate with other small districts and share the cost. We will need to work out a contract to fit your specific situation.

To reach me:

- email jim@prealgebra.net or jslosson@aol.com (I hope to be the last subscriber.)
- Snail mail: Jim Slosson—Math Lab, 5515 Cedar Flats Road SW, Olympia, WA, 98512

About the Author

Jim (James) Slosson started teaching printing—photography and yearbook— in 1973. He was the worst beginning teacher he ever met. Yet he received the Washington State Christa McAuliffe Award for excellence in teaching just eighteen years later. He learned to teach by trying to improve every single day of his fifty-year career.

After a stint fixing Puget Sound Alternative High School, possibly the worst high school in Washington State, Jim retired to write Math Lab/AGA and spent eight years developing the program in Elton School District.

Now in his midseventies, Jim spends his time substitute teaching, raising grandchildren, and sailing.

Jim Slosson has authored over fifty articles and several books. If he lives long enough, he plans to finish his book *How to Fix a Broken School.* He has an outline ready for *Couldn't Find Their Own Asses with Both Hands in a Well-Lit Room—Why Education Can't Fix Itself.*

Other works by Jim Slosson:

Unmailed Letters from a REMF, a realistic and raunchy history of the war in Vietnam from a clerk's perspective. It accidentally turned into a novel.

Over sixty YouTube videos covering the operation and maintenance of small sailboats. Jim's channel is called Jim's Little Boat.

About the Author

Glossary

ACT: American College Test.

AGA: Algebra/Geometry/Arithmetic. See Math Lab.

ASVAB: Armed Services Vocational Aptitude Battery (Some states allow ASVAB scores in math to meet state graduation requirements.)

Begin with the end in mind: See Stephen Covey, one of the key principles of The 7 Habits of Highly Effective People.

Canvas: A cloud-based learning management system (LMS) specifically designed for K-5 on up through higher education institutions.

Covey – Stephen Covey: American educator and author. His most popular book is *The 7 Habits of Highly Effective People.*

CYA: Cover your ass—I learned it in the army.

Google Classroom: A Google application for classroom management and communication.

Formative Assessment: Long, long definition—the quiz.

Guskey—Thomas Guskey: Internationally known expert in evaluation design, analysis, and educational reform who championed "mastery learning."

Hattie—John Hattie: Professor of Education from New Zealand and a key proponent of evidence-based teaching. Dr. Hattie is a very smart guy, but he has never actually taught school.

Kahn Academy: Khan Academy is a nonprofit with the mission of providing a free education for math, and other academic areas.

MAP: Measures of Academic Progress is an assessment system (test).

Marzano—Dr. Robert Marzano: An educational researcher in the United States. He has done educational research and theory on the topics of standards-based assessment, cognition, high-yield teaching strategies, and school leadership.

Mastery Learning: See Thomas Guskey.

Math Lab/AGA: An instructional design including materials developed by James Slosson in 2003–2022.

McLogan—Brian McLogan: American teacher who is famous for his self-titled YouTube channel. He has made more than 12,000 math videos.

Meyers–Briggs: Katharine Cook Briggs and Isabel Briggs Myers researchers in the understanding of psychological type.

Non-homework: Work done in class under teacher supervision.

SAT: Scholastic Assessment Test, an entrance exam used by many colleges and universities to make admissions decisions.

PBIS:, a collection of good ideas to improve school climate and discipline—often over-used.

Rolling Review: Curriculum design that emphasizes a review of previously learned concepts throughout the entire course.

Saxon Math: Math curriculum in which students learn incrementally, one lesson at a time, using knowledge acquired in previous lessons. See rolling review.

SBA: Smarter Balanced Assessment—a comprehensive assessment system used by many state agencies and individual school systems.

Skyward: A school management software system.

Smarty Pants: Kids in your class who know every answer.

Summative Assessment: The test—usually, the final.

Wooden—John Wooden: American basketball coach and motivational speaker famous for his tenure at UCLA.

TESA: Teacher Expectations and Student Achievement—was developed in the 1970s by the Los Angeles County Office of Education.

Appendix A

Mr. Slosson's Standards for Highly Employable Young Adults

Students will demonstrate the qualities of young adults preparing themselves to be productive members of the workforce and the community. We are working on the behaviors of highly employable young adults so that they become habits.

1. *Call the teacher by his surname, "Mr. Slosson."*
2. *Start and end every class seated at your workstation. That is how we take roll, engage in a community-building activity, and start the day's lesson.*
3. *Have your binder out on your table and ready. Make sure you have a sharp pencil.*
4. *Be courteous. Say, "Please, thank you, and you're welcome."*
5. *Say, "Excuse me," if you need someone to move or if you bump into them.*
6. *Maintain a pleasant countenance. Do not "stink eye" other people.*
7. *Listen with your ears and your body language. Lean into the conversation.*
8. *Obey all legal requests from your teacher. Do not expect a detailed explanation.*
9. *Hand things and accept them nicely. Do not grab. Do not throw, toss, lob, or otherwise cause objects to become airborne. Zero trajectory.*
10. *Use professional office workplace language. Do not use demeaning language. Do not make comments about people's race, sexuality, or social circumstances. Do not swear or curse. Use a conversational voice.*
11. *If someone treats you with disrespect, treat them with respect anyway.*
12. *Do your makeup in the bathroom or outside of class. Do not do your hair or make-up inside of class.*

13. *If you must come in late, quietly enter the room, get your binder, and find a seat. Do not make a commotion or tell everyone why you are late.*
14. *Use "I" messages instead of "you" messages, i.e., Don't "should on other people."*
15. *Comply with requests. Do not argue or explain. A simple nod or "okay" is sufficient.*
16. *Sit in your assigned seat.*
17. *If you are asked to move, do it quickly and quietly. If you want an explanation, ask later.*
18. *Follow our procedure to go to the "restroom" or "lavatory." Do not engage in detailed descriptions of the things you will do when you get there.*
19. *If you need to answer the telephone, say "D7, student speaking."*
20. *Keep your phone face down on your desk or keep it in your pocket per classroom rules.*
21. *Help other people with their work, but do not "copy across."*
22. *Check with two students before asking for teacher help.*
23. *Do all your work in pencil, unless there is an unusual circumstance. (If work is done in pen, it will be returned ungraded.)*
24. *Sit in chairs properly. Do not sit on tables, the backs of chairs, or other furniture.*
25. *Work at your assigned seat and get up to move around only if it is on "math business."*
26. *Clean up your own things and a little bit more.*
27. *Attend to the math work. No outside reading or devices unless you are caught up.*
28. *Respect our equipment and furniture. Do not abuse rulers, calculators, or other equipment. Do not write on color, or deface furniture. Do not write on binders.*
29. *Keep your work in class. Homework is a privilege, not a right.*
30. *If you don't like a request, do it anyway and complain to the teacher or administrator later.*

Math Lab Standards for Highly Employable Young Adults Quiz

Quiz for (name) ______________________________ *Date:* ________ *Per* ______

1. What is the appropriate form of address for the teacher?

____

2. How do we start and end every period? ______________________________

3. Which way do we face for lessons? ______________________________

5. When is it okay to listen to music/play games? ______________________________

6. It's not okay to copy across, but okay to: ______________________________

7. When may you use your device? ______________________________

8. How do we give things to other people? ______________________________

9. What are the courtesy words? ______________________________

10. Where do we sit for labs or activities? ______________________________

11. How many people at a table for lessons? ______________________________

12. What color pen should be used for math work? ______________________________

13. How do we answer the telephone? ______________________________

14. What is the appropriate response for instructions? ______________________________

15. When is it okay to argue with the teacher? ______________________________

16. When is it okay to throw, toss, lob, etc.? ______________________________

17. How many people do you "check with?" ______________________________

18. What are the rules for rulers? ______________________________

19. What kinds of language are not allowed? ______________________

____

20. Where do we start and end every class? _______________________

21. When is it okay to talk on a device in the class _______________

22. What do you do if you need to take an emergency phone call?

____

Appendix B – Math Lab/AGA interview activity

Interviews — This activity provides a chance for students to learn each others names, build some group cohesion and practice group process skills appropriate for young adults.

The class sits in a circle. One student is interviewed; one student is the moderator. Each student has two numbered cards that have an interview question. The moderator calls out the cards in order and the students read the question.

The rules: The students must focus on the interviewee. The student with the question can ask a follow-up question, but cannot argue with the interviewee. The interviewee can pass on a particular question. Students can ask follow-up questions when the interview is over.

Nice touch. Each person asking a question should start with the interviewee's name. The questions can be written out on 3x5 index cards and laminated. It usually works best if you make sure that students do not have cards that are too close together in the numbered sequence.

After the interview is finished, ask each student to tell one thing they remember about the student who was interviewed.

Some possible questions.

1. What is your full name?
2. Were you named after anyone?
3. How old are you?
4. Which part of the district do you live in?
5. Do you have any assigned chores at home?
6. Do you have siblings? How old? Names?
7. How do you get along with your siblings?
8. Do you have any pets? (follow-up)
9. Do you have chores at home?
10. Do you have a job (even informal or part time)?
11. What was your favorite toy when you were little?
12. Who was your best friend when you were under 10 yrs old?
13. What is your favorite kind of music?
14. Where would you like to live when you are an adult?
15. What kind of car would you like to own?
16. What kind of career job would you like to have?
17. Do you plan to attend some kind of school after high school?
18. Who is your favorite relative that doesn't live with you?
19. If you could meet a celebrity, who would you pick?
20. What is your favorite dessert?
21. What food do you like the least?
22. What is your favorite class?
23. What is your least favorite class?
24. Would you come to school if you did not have to?
25. What kinds of things are a struggle for you?
26. What kinds of things are easy for you? (What are you especially good at?)
27. What are you afraid of?
28. Can you think of anyone you admire?
29. If you could change one thing about school, what would that be?
30. If you could change one thing about yourself, what would that be?
31. What kind of clothes do you like to wear?
32. What is your favorite kind of music?
33. What is your favorite movie?
34. What is your favorite television show?
35. What is your favorite holiday?
36. Do you think you might join the military?
37. How many kids would you like to have?
38. Did you have a special teacher in elementary?
39. How do you decide when you are successful?

Your most powerful lesson—stop at Q. ______

Appendix C – the one-minute multiplication test

5	8	7	9	5
X5	X6	X6	X8	X9
7	6	9	6	6
X7	X5	X6	X9	X7
5	8	7	5	8
X8	X7	X5	X6	X8
6	7	8	9	6
X8	X9	X5	X7	X6
8	5	9	7	9
X9	X7	X9	X8	X5

Appendix D – sample Math Lab/ AGA computational assignment

Name ______________________ Date ________ Period ________

Math Lab, Lesson 31
Add/Sug Neg #, Add/Sub Fractions, Probability, Discounts, Ratios, Averages
Measure Twice. Cut Once.

-3 0 3

1. Fill in the number line.

2. Counting numbers are like integers except integers include all the positive numbers, the ________ numbers, and ______.

3. $3 + 3 =$ ____, $3 - 3 =$ ____, $-3 - 3 =$ ____, $-3 + 3 =$ ____, $-7 + 10 =$ ____, $7 + 6 =$ ____, $13 - 10 =$ ____

4. $5 + (-2) =$ ____, $(-2) + (7) =$ ____, $-2 - 2 =$ ____

5. Another way to say "subtract" is to say ______ a negative ______.

7. Addition—subtraction ________ action.

8. $\frac{1}{2} + \frac{1}{4} =$ ____, $\frac{1}{4} + \frac{1}{3} =$ ____, $\frac{1}{8} + \frac{1}{3} =$ ____

 Lesson 31, Page 1

Name ______________________________ Date __________ Period __________

9. $\frac{1}{a} + \frac{1}{2} =$ ____, $\frac{b}{2} + \frac{1}{3} =$ ____, $\frac{1}{8} + \frac{1}{4} =$ ____

10. $\frac{1}{3} \cdot \frac{3}{5} =$ ____, $\frac{1}{4} \cdot \frac{2}{3} =$ ____, $\frac{3}{5} \cdot \frac{7}{8} =$ ____

11. $\frac{1}{a} \cdot \frac{3}{5} =$ ____, $\frac{b}{4} \cdot \frac{2}{3} =$ ____, $\frac{c}{d} \cdot \frac{7}{8} =$ ____

12. $.04\overline{)16.4} =$ ____ $\frac{.08}{2} =$ ____

13. $\frac{3}{2} \div \frac{1}{2} =$ ____, $\frac{1}{2} \div \frac{3}{2} =$ ____

14. 4' 6' y x 9'

A= ________ P= ________

15. How did you find the length of x?

Name ____________ Date ________ Period ________

16.

7'

d= ____
r= ____
C= ____
A= ____
π = 3

A = 48 sq"
$A = \pi r^2$
$48 = 3 \cdot r^2$
$\frac{48}{3} = \frac{3 \cdot r^2}{3}$

r= ____ d= ____
C= ____

17.

a

b

∠a = ____
∠b = ____

18.

10mi

6mi

You need to find the length of Slume Lake but you can't measure it directly. You can step off the lengths shown in the drawing. How long is the lake? Show all work.

length: ____

Name ______________________ Date __________ Period __________

19. Coats cost \$28, \$75, \$112, \$80, \$80
Find the measures of ________ tendency.

mean = ______
mode = ______
median = ______
range = ______

20. If a person asked "About how much do coats cost?" What would you answer? Why?

cost = ______

I would say that because

21. How much is the \$80 coat if it is discounted 35%? How much w/ 10% tax

orig cost ______
cost w/ disc ______
tax ______
new total ______

22. Show as decimals.

1% = ______, 10% = ______ 100% = ______ 25% = ______

$\frac{1}{4}$ = ______, $\frac{1}{2}$ = ______ $\frac{1}{3}$ = ______ $\frac{2}{3}$ = ______

Appendix E – sample Math Lab/AGA assignment

Math Lab
$30°/60°/90°$ Triangles
(estimated time 1 hour)

Measure Twice. Cut Once.

Summary of the activity: You will draw some right triangles that have a hypotenuse twice as long as the shortest leg. Then you will measure the angles with a protractor and fill in some questions. You will reach a conclusion about 30/60 right triangles.

Point value: 10

Words to know: Define the following words.

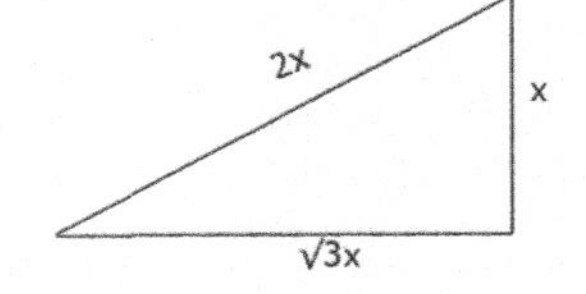

1. Right Triangle
2. Hypotenuse
3. Leg of a triangle
4. Protractor
5. Opposite angle: Use the illustration to show angle opposite the short leg, label it a. Use the illustration to show the angle opposite the hypotenuse, label it b.

Materials: Pencil, ruler, protractor.

Procedure:

1. Make sure that you can measure lines with a ruler and angles with a protractor.
2. Measure the short leg of Triangle A on page 3. How long is it? _________"
3. How long is two times (x 2) the short leg? _________"
4. Draw in the hypotenuse of Triangle A so that the hypotenuse is twice as long as the short leg.
5. Draw in the long leg of Triangle A with a heavy line so that you have a neat, easy to see triangle.
6. Now draw in Triangle B. How long is the short leg?

   _________"

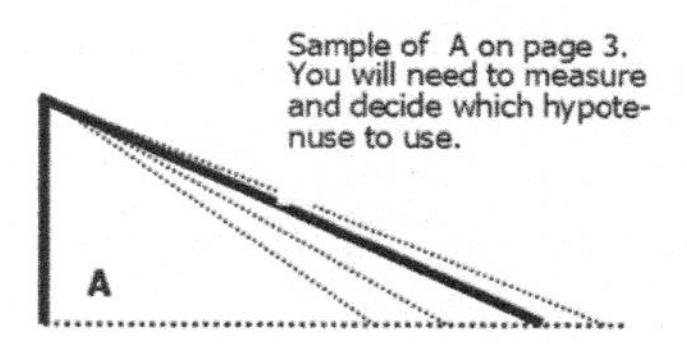

Sample of A on page 3. You will need to measure and decide which hypotenuse to use.

7. Draw in the hypotenuse of Triangle B so that it is twice as long as the short leg.
8. Draw in the long leg of Triangle B.
9. Now draw Triangle C. How long is the short leg of Triangle C?

   ___________"
10. Draw in the hypotenuse of Triangle C so that it is twice as long as the short leg.
11. Draw in the long leg of Triangle C.
12. Now draw Triangle D. How long is the short leg of triangle D?

   ___________"
13. Draw in the hypotenuse of Triangle D so that it is twice as long as the short leg.
14. Draw in the long leg of Triangle D.
15. Now use the protractor and measure the angle opposite the short leg. That will be the more acute (sharpest) angle. Record your results for each triangle. Remember to use the degree label.

D

C

A

B

sample of page 3

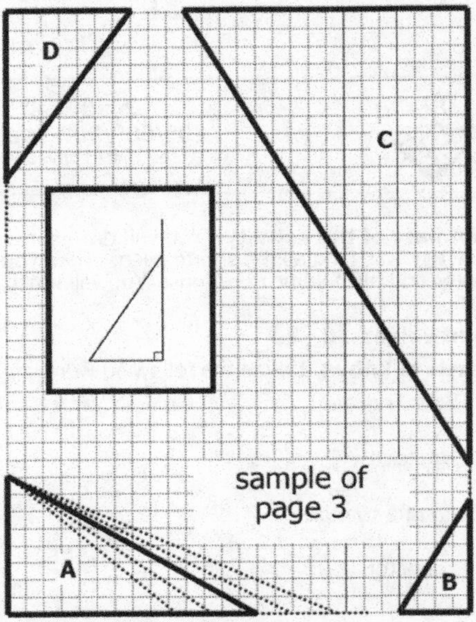

16. Measure of angle opposite the short leg for the following triangles:

   A _________ B _________ C _________ D _________
17. Now use the protractor and measure the angle opposite the long leg. That will be the less acute angle. Record your results for each triangle; remember to use the degree label.
18. Measure of angle opposite the long leg for the following triangles:

   A _________ B _________ C _________ D _________
19. In every case what is the measure of the angle opposite the hypotenuse?

   ____

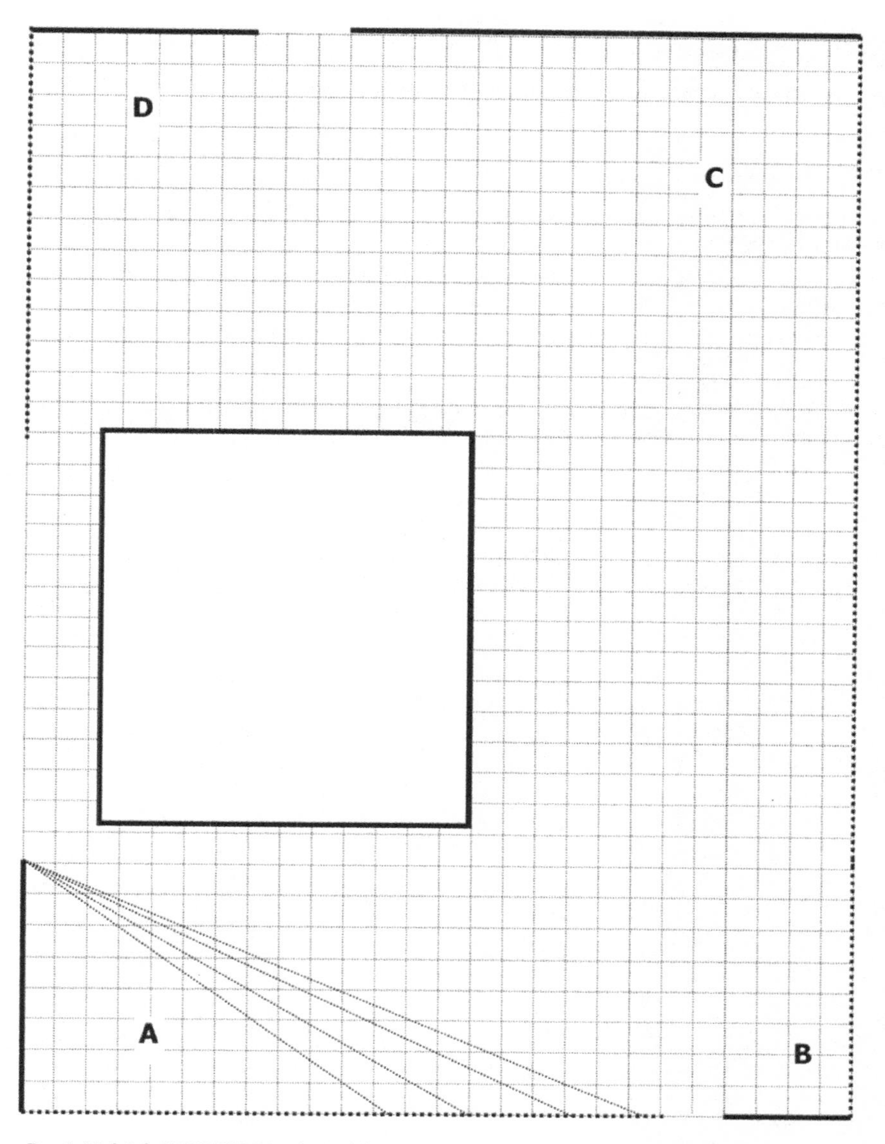

Page 3, Math Lab, 30°/60°/90° Triangles Activity

20. Fill in the following blanks.

If one of the angles of a right triangle is 30,° the other angle must be ________.

If one of the angles of a right triangle is 60,° the other angle must be ________.

If two of the angles of a triangle are 60° & 30,° the third angle must be ________.

If a triangle is a 30/60 right triangle, the hypotenuse must be ________ times longer than the short leg.

If a triangle is a 30/60 right triangle, the short leg must be ________ as long as the hypotenuse.

21. In the blank space on page three draw a 30/60/90 right triangle. Make the short leg 3 centimeters. (3 com). Label the following: 30° angle, 60° angle, 90° angle, hypotenuse, short leg, long leg.

22. How would these rules change if you used centimeters to measure instead of inches? Why? (Write at least two complete sentences.)

Made in the USA
Coppell, TX
29 September 2022